5.-

IVORY AND ITS USES

IVORY
AND
ITS USES

by BENJAMIN BURACK

CHARLES E. TUTTLE COMPANY
Rutland · Vermont : Tokyo · Japan

REPRESENTATIVES

Continental Europe: BOXERBOOKS, INC., *Zurich*
British Isles: PRENTICE-HALL INTERNATIONAL, INC., *London*
Australasia: BOOK WISE (AUSTRALIA) PTY. LTD., *Sydney*
104-108 Sussex Street, Sydney 2000

Published by the Charles E. Tuttle Company, Inc.
of Rutland, Vermont & Tokyo, Japan
with editorial offices at
Suido 1-chome, 2-6, Bunkyo-ku, Tokyo

Copyright in Japan, 1984
by Charles E. Tuttle Co., Inc.

All rights reserved

Library of Congress Catalog Card No. 83-51417
International Standard Book No. 0-8048 1483-x

First printing, 1984

PRINTED IN JAPAN

TABLE OF CONTENTS

LIST OF ILLUSTRATIONS 9

PREFACE 11

PART ONE : HISTORY, ART, AND CRAFT

I : Historical Development 15
    Early Ivory Craft   16
    Ivory in Asia   16
    European Ivory Craft   19
    Ivory Trade and Carving in Africa   23
    Ivory in North America   26

II : Sources of Ivory 28
    Elephant   28
    Fossil: Mastodon and Mammoth   30
    Walrus   31
    Narwhal   32
    Hippopotamus   33
    Boar and Warthog   34
    Sperm-Whale Teeth   34

Table of Contents

III : Ivory Substitutes 35
    Bone 35
    Horn 36
    Hornbill 37
    Vegetable Ivory 38
    Artificial Ivory 39

IV : Cutting and Carving 41
    Preparation of the Ivory 41
    Attempts at Sheet-Ivory Production 43
    Mechanical Cutting Methods 45
    Ivory Carving as a Hobby 47

V : Coloration of Ivory 49
    Dating and Fake Aging 49
    Artistic Coloration 51

VI : Care and Cleaning 53

VII : Testing for Ivory 56
    Specific Gravity Test 56
    Chemical Tests 57
    Burning Tests 57
    Microscopic Test 58
    Light-Reflection Test 58
    Simple Tests 59

PART TWO : THE USES OF IVORY

1. Artistic Ivories 77
2. Books and Book Accessories 89
3. Clocks, Clock Cases, and Accessories 91
4. Cloth and Clothes Making, Embellishment, and Repair 92
5. Containers: Boxes, Cases, Holders 95
6. Costume Articles and Accessories 103

7. Educational Articles 111
8. Food-Related Articles 113
9. Furniture and Home Furnishings 118
10. Games and Game Accessories 122
11. Health Articles and Accessories 130
12. House Structural Accessories 135
13. Musical Instruments and Accessories 136
14. Personal Comfort Articles 139
15. Personal Grooming Articles 160
16. Puzzles 165
17. Religious Articles 166
18. Scientific Instruments and Aids 177
19. Smoking Articles and Accessories 178
20. Toys 181
21. Weapons, Hunting and Fishing Articles, and Accessories 183
22. Writing and Painting Instruments, and Related Articles 187
23. Miscellaneous 194

APPENDIX: Museum Collections 205

NOTES 209

BIBLIOGRAPHY 225

INDEX 229

LIST OF ILLUSTRATIONS

(Asterisk indicates color plate)

*1. Vegetable-ivory products 65
*2. Persian miniature painted on ivory 65
*3. Decorative stand 66
*4. Bracelet-and-earrings set 66
*5. Netsuke 66
6. Grainy appearance of some ivory 67
7. Cheverton's reproducing machine 67
8. Artificial ivory products 68
9. Artistic ivories 68
10. Erotic die 68
11. Ivory curiosities 69
12. Miniature carvings 69
13. Whale-tooth scrimshaw work 70
14. Puzzle ball 70
15. Magnifying lens and book-stand 70
16. Bookmarks 71
17. Book covers 71
18. Clock accessories 71
19. Hooks and needles 72
20. Cloth and clothes making, and repair articles 72
21. Pair of cases and sectioned container 72
22. Purse frames, snuff bottles, and snuff box 73
23. Box and round container 73
24. Vase 74
25. Candle holder 74
26. Set of necklace, bracelet, and earrings 74
27. Ivory bracelets 74
*28. Calling-card cases 75
*29. Clothes making and repair accessories 75
*30. Plaque bust of D. Eisenhower 75
*31. Food-related articles 76
*32. Health articles 76
*33. Game articles 141
*34. Chinese puzzles 142
*35. Ivory-handled walking sticks 142
36–38. Costume articles and accessories 143, 144

[9]

39. Models of the human eye *144*
40. Jewelry case *145*
41. Teapoy from India *145*
42. Food-related articles *145*
43. Game-scoring devices *146*
44. Alphabet set and globe *146*
45. Game articles *146*
46. House structural accessories *147*
47. Batons and parts of musical instruments *147*
48. Fans and tweezers *148*
49. Personal comfort articles *148*
50, 51. Personal grooming articles *149*
52. Letter openers and paper creaser *150*
53. Toy fish and whistle *150*
54. Religious articles *150*
*55. Daggers *151*
*56. Cigarette and cigar containers *151*
*57. Smoking articles and accessories *152*
58. Articles related to weapons, hunting, and fishing *153*
59, 60. Articles related to writing *153, 154*
61. Gavel, sled runner, and baton *154*
62. Handles, boat spur, and tools *155*
63. Whisk broom and pick, knife, corkscrew, and ruler *155*
64. Miniature Masonic tools *156*
65. Sewing kit, latch peg, and ivory coin *156*
66. Ivory products of uncertain use *156*

PREFACE

This book was preceded by a lifelong hobby of collecting ivory artistic carvings and useful products, many of which are illustrated here. They were gathered from various sources: antique and curio shops in the United States, Europe, Asia, Africa, and the Middle East; auction sales; "For Sale" listings in hobby magazines; my own "Ivories Wanted" advertisements in such magazines; and friends who offered to sell or give me some ivory article they no longer wanted.

Teeth are the hardest substance in the body, and tusks are a specialized form of teeth, overgrown and usually extending beyond the mouth. The main component of tusks is ivory, and this feature led to a paradoxical situation. Mankind has utilized a great variety of products made from ivory, in addition to using the meat, hide, and oil from the carcasses of the animals from which the substance was taken. But counterbalancing that advantage were two disturbing consequences. One was the nineteenth-century African slave trade that became the story of "black and white gold"—the enslavement of African blacks for carrying the heavy white tusks of elephants. The second consequence was the increased slaughter of the major tusked species, to the point where some species came close to extinction. The latter problem has yet to be fully resolved.

Part One will show that for at least five thousand years man has been fascinated by ivory; and for us now, ivory art and artifacts reveal the myths, customs, dress, work, amusements, and other features of the domestic life of the past. Part Two is probably the first encyclopedia of ivory uses; it will be of special interest and value to collectors, dealers, libraries, and museum curators. The Appendix offers an extensive list of worldwide museum collections of ivories; this will be of interest to collectors and any readers who in their travels may want to see ivory art and utilitarian products.

With the two exceptions cited below, all photographs in this book are by Wayne McCall of Santa Barbara, California. Figure 7, showing a reproducing machine, is by courtesy of the Science Museum, London; Figure 39, showing models of the human eye, is by courtesy of the German National Museum, Nuremburg.

—BENJAMIN BURACK

PART ONE

HISTORY, ART, AND CRAFT

Note: All ancient datings cited in this work are approximate. Metric equivalents given in parentheses after measures are provided as a general guide for the convenience of readers. They do not represent precise conversions. (1 inch equals 2.54 centimeters.)

I

HISTORICAL DEVELOPMENT

———◆———

Ancient carvers of ivory may not have realized how durable their works would be. Some surviving examples have endured for thousands of years, outlasting most of the ancient works of art that used such materials as paper, cloth, or wood.

The great appeal of ivory has been its clean beauty, smooth and soft feel, power to show a bright gloss, and capacity to accept coloring for artistic embellishment. It can be cut, penetrated, filed down, and then carved in elaborate detail. Its subtle color led ancient poets and other writers to allude to ivory when referring to healthy skin tone. It is no wonder that ivory was preferred for artistic work, while bone, its closest substitute, was used mainly for utilitarian products.

There was little or no waste. Small bits of residual ivory were used as design inlays in furniture, boxes, musical instruments, and weapons. Even the residual ivory dust has been highly prized. At one time it was thought to be an effective medicine, and much later it was used in burnt form for making india ink and the oil color ivory black. The waste dust has been used in a hardening compound for ivory figurines, and also as size for filling the porous surface of paper and cloth.

Because of the great durability of ivory, its history can be traced back to the beginnings of civilization. Prehistoric remains in Asian and European caves have yielded tusk and bone fragments that show incised sketches of mammoths, as well as engravings of fish, snakes, wild goats, and reindeer.

Ivory became associated with gold and silver as a luxurious commodity, used especially for decorating temples, furniture, and objects of value. The Old Testament includes references to ivory used in this way. One passage tells that King Solomon "made a great throne of ivory and overlaid it with pure gold" (1 Kings 10:18, and 2 Chronicles 9:17). More strikingly, Ahab allegedly made a house of ivory (1 Kings 22:39). But lest it seem that ivory was always honored, we can note a dark and prophetic reference too: "And the houses of ivory shall perish" (Amos 3:15).

EARLY IVORY CRAFT

The earliest surviving ivory carvings are mainly from Egypt, Phoenicia, Assyria, China, and, somewhat later, India. In predynastic Egypt (before 3100 B.C.), tombs contained ivory and bone fishhooks, arrowheads, pins, bodkins, spoons, knives, combs, bracelets, and necklaces.[1] Early Egyptian graves have also yielded ivory hairpins, finger rings, armlets, anklets, mirror handles, vases, spatulas, play marbles, game boards with dice, and statuettes.[2] Tombs of the early dynasties' kings included ivory labels on jars of balsam-tree oil, ivory playing-pieces for games, and ivory inlay-sections in beds, furniture, and gambling tables.

In nearby Phoenicia, craftsmen developed an ivory-carving industry that flourished there and in Syria from the sixteenth century B.C. until the seventh century B.C. Prized articles included ivory-inlaid furniture such as couches, chairs, and tables. Phoenician craftsmen traveled widely and may have fashioned the ivories of Phoenician style excavated at ancient Megiddo in Palestine (dated between 1350 and 1150 B.C.). These ivory items include handles, a box, a pen case, jar lids, shallow bowls, spoons in the shape of nude females, combs (some with teeth along both edges), game boards and markers, furniture ornamentation, medallions, artistic plaques, nude female figurines, and small carvings of animal heads.[3] Another reference reports additional ivories found there—vases, ointment containers, and part of a lyre.[4]

The British Museum has probably the best collection of the Assyrian ivories (1200–600 B.C.), excavated by Austen Henry Layard from 1845 to 1851 at the presumed site of ancient Nineveh, later called Nimrud. Ultimately uncovered were ivory pictorial panels, figurines, combs, containers, handles (including one on a fly whisk), knobs, a spoon, an apparent inkwell, and numerous other ivory articles. The digging continued off and on for a hundred years. After 1926, more and more of the remaining fragments went to the area's own museum in Baghdad, Iraq. There seems little hope of ever joining them to the related fragments in the British Museum.[5]

IVORY IN ASIA

Elephant tusks were used throughout ancient China; in addition, walrus tusks were utilized in the north, and hippopotamus tusks in the warm southern regions. When its own sources of elephant tusk became insufficient, China imported the tusks from Southeast Asia and India. The first Chinese dynasty (Shang, 1500–1027 B.C.), was already producing ivory animal figures, knife handles, sword hilts, combs, hairpins, and decorated pail-shaped vessels made from the widest section of the elephant tusk.[6]

Then the Chou dynasty (1027–256 B.C.) furnished ivory chopsticks, finger rings, and pins. During this time, decorative ivory tips were attached to the ends of the bow used in hunting, emperors' chariots were embellished with ivory pieces, and important personages sometimes carried ivory writing-tablets suspended from their waist girdle. A richly carved ivory tool for unraveling knots, made in this period, is now at the Metroplitan Museum of Art in New York. After a brief decline in the popularity of ivory articles, the Han dynasty (206 B.C.–A.D. 221) produced ivory seals for officials, personal ivory batons for dignitaries, and then body armor in the form of overlapping walrus-ivory plates (much lighter than metal armor and more enduring than leather) for the Su-shen tribesmen of northeast Asia.[7]

Following a longer period of uncertainty for ivory carving, the aristocratic class of the Tang dynasty (A.D. 618–906) began to favor the use of the material for their fine articles; these included jewelry boxes, small folding screens, flutes, plectrums for stringed musical instruments, game pieces, hairpins, combs, and measuring rulers. In later periods, some of these articles continued to be produced; but other items of ivory were introduced, including knife handles, sword hilts, ornaments, inlaid furniture, back scratchers, fans, cigarette holders, scent boxes, incense boxes, Mah Jong game-pieces, and writing accessories such as arm rests, paintbrush handles, paintbrush cases, and seal boxes.

It was not until the Ming dynasty (1368–1644) that emphasis was placed upon carving purely artistic ivories.[8] Now, beautiful human figurines were carved for palaces, temples, and richly built homes. The last dynasty, Ching (1644–1911), further popularized the charming "doctor's lady" figure, an ivory female nude designed for use by modest ladies visiting their doctor; they could merely point to the part of the reclining figure that corresponded with the location of their ailment. The fascinating puzzle balls that were carved one within another also continued to be made throughout this time. Other ivory articles included fans, calling-card cases, chess sets, cigarette holders, and snuff bottles, as well as miniature temples, pagodas, houses, and pleasure boats.

An exquisite example of artistic skill may be noted in the following description of a ceremonial fan at the Field Museum of Natural History in Chicago: "A marvel of technical skill and harmonious beauty, being plaited from finely cut ivory threads held by a tortoise-shell rim and overlaid with colored ivory carvings of lilies, peonies, asters, and a butterfly. The handle, likewise of ivory, bears etched designs in colors, of flowers and butterflies."[9]

In Japan, ivory carving began around the eighth century A.D., and was influenced by the ivory craft of neighboring China, from which the tusks were imported. During the following centuries, favorite ivory articles included small boxes, pipe cases, needle cases, sword sheaths, fans, small replicas of shrines, and somewhat later, small figurines, or *netsuke* (see p. 81). Carvings also included ivory flutes and rulers. Sen-rei Soma wrote the first Japanese manual on the subject of ivory craft, entitled *The Methods of Ivory Carving*; it has brief introductions written by distinguished Japanese.[10] The variety of tools used by Japanese carvers of ivory appears to be greater than those used by ivory craftsmen of most other countries. Modern utilitarian ivory articles have included chopsticks, chopstick holders, cocktail picks, table coasters, orange peelers, and shoehorns.

There is evidence that ivory craft in India was already well developed before the second century B.C. By the first century A.D., ivory products included figurines, combs, and handles. Chess pieces appeared during the eighth century. By the seventeenth century, Ceylon was furnishing dagger handles, archers' thumb-guards, fencing-foil guards, powder horns, fan handles, water dippers, book covers, dice, potters' dies, compasses (dividers), pill boxes, small musical drums, and coffers.[11]

Edgar Thurston, former Superintendent of the Madras Government Museum, has related the interesting history of India's ivory industry of the 1800s. When one maharajah, Rama Varma, saw some fine ivory carvings, he became enthusiastic about this craft and began to strongly encourage a broader application. The maharajah who succeeded him, Marthanda Varma, sent to Queen Victoria a new ivory throne, elaborately carved, and decorated with inset jewels. After prominent display at the Great Exhibition of 1851 in London (the first World's Fair), it was installed in the State Apartment at Windsor Castle. Nineteenth-century ivory products included individual cases for calling cards, envelopes, gloves, handkerchieves, jewelry, money, and stamps. There were also bookstands, chessboards, inkstands, letter openers (paper knives), paperweights, penholders, pen racks, picture frames, rulers, teapoys (small three-legged tables), walking sticks, and watchstands.[12]

Also carved in Madras were scent bottles, eye-paint containers, special talismans for children, ornamentation for musical instruments, and howdahs (seats) for carrying royalty on the backs of elephants.[13] Especially in western India, ladies' combs and arm bangles were very common ivory products. The residual ivory sawdust (produced when sections are cut from a tusk) was sometimes sold to cow and buffalo dealers, who fed it to the female animals in the hope of increasing their milk yield. In northern India, ivory sawdust was considered to be a fortifying medicine.

As in some other countries, ivory carving in India sustained itself by way of a family line. A boy might start such work at the age of ten; after a few years of training, he would be paid a little when one of his ivory carvings was bought by someone. In ten to twenty years, he could become a master craftsman.[14]

Considering the elephant's usefulness for labor in the forests while it was alive, and the use of its ivory after it died, it is not surprising to learn of an old Indian proverb: "In life and death, the elephant is worth a thousand pagodas."

The Islamic world had begun developing its own ivory craft by the eighth century A.D. Featured were small ivory cases, some with locks and hinges of gilded silver and gold. The following centuries saw ivory hunting-horns, richly adorned ivory handles for daggers and scimitars, and then powder horns, after gunpowder and firearms had been invented. Recent centuries featured toiletry articles and also unusual articles such as ivory inkwells. Almost uniquely Islamic are the fine wooden jewelry boxes, tables, and doors, which have been so intricately inlaid with tiny ivory pieces that they are truly masterpieces of craftsmanship.

EUROPEAN IVORY CRAFT

In Greece, ivory craft had already been established by 1000 B.C., possibly earlier. It then seems to have died out for a period, but was revived in the eighth century B.C. The earliest surviving carvings appear to be the ones of small female figures, thought to have been used in votive practices. For several centuries, Greece was a major center for ivory carving. The combining of styles from East and West can be noted in early Greek vases, plaques, and statuettes of ivory.

Around 560 B.C., Smilis carved a set of ivory statues representing the Hours of the day seated on thrones. The great fifth-century sculptor Phidias fashioned massive statues that had ivory surfaces for the face and hands; the goddess Athena stood nearly 40 feet (12 m.) high in the Parthenon at Athens; and the god Zeus was about 58 feet (18 m.) high at Olympia. Ivory statues, perhaps representing the deities, adorned the ship of the Macedonian ruler of Egypt, Ptolemy II (Philadelphus), who reigned from 285 to 246 B.C.[15] Pausanias, the second-century Greek geographer-historian, wrote about large ivory statues of his time—Amphitrite, Poseidon, Hera, Endymion, and others—some of them overlaid with gold leaf.

We may note that, lacking a secondary use as a substance, ivory did not offer the same commercial temptation to potential destroyers as did the great works made of precious metals. Most of the gold and silver productions were later melted down into salable bullion, and thus disappeared forever as art. Marble statues were often burned down in order to make lime, and countless fine wooden carvings became the fuel that provided warmth at some later time.

Another providential aspect for ivory was that most carvings, being relatively small and portable, could be hidden; thus, they usually escaped the ravages of religious fanaticism that defaced and disfigured so many superlative religious representations in wood and marble. That large works of ivory could not escape this scourge is indicated by the fact that the great ivory statues in Greece were probably destroyed by Christian fanatics of later times. Similarly, much Christian ivory art was later destroyed by fanatics of other religions.

In Roman times, the fascination with ivory extended to decorating saddles and chariots with small inlay pieces. The historian Livy tells that Romans celebrated Lucius Scipio's great victory in Asia Minor (190 B.C.) by parading through the streets of Rome with a captured loot of more than 1,200 elephant tusks.[16] The Roman Empire had ivory thrones, chairs, door panels, boxes, dice, musical instruments, small animal cages, baskets, combs, scepters, shields (made of ivory sections), statuettes, and other articles. But ivory adornment had Roman detractors (as among biblical prophets, earlier). In Juvenal's *Satires* of the early second century A.D., we note his bitter comment about rich Senators' luxurious use of ivory for couches and tables. Earlier, Propertius, in his *Elegies* (10 B.C.), had lamented the desire of certain personages for ivory bedposts and chariot doors; he felt especially repelled by a temple of ivory.[17]

After Emperor Constantine chose Byzantium as his new capital city in A.D. 330 (later renamed Constantinople after himself), it became a major center for ivory carving. The Byzantine style of ivory work spread westward to Italy, Spain, France, and other Mediterranean countries. When the empire split into two separate sovereignties in 395, the Eastern Roman Empire experienced a magnificent development in architecture and art, including ivory craft, which continued for over a thousand years until Constantinople fell to the Turks in 1453.

With the coming of Christianity to the Roman World, many ivory carvers turned away from pagan inspirations and began to fashion ivory objects related to worship. The main centers of this work were Alexandria in Egypt, Antioch in Syria, and, most importantly, Constantinople. For example, pyx boxes, formerly for any contents and which had depicted games and hunting scenes on their surfaces, now contained the sacrament and portrayed religious scenes. Writing tablets, earlier established as gifts from newly appointed Roman consuls, now became ecclesiastical recording tablets.

Beginning with the reign of Eastern Roman Emperor Leo III in the early 700s, the creation of images was suppressed, especially after his official edict of 726. This Act started a wave of destruction of sculpture of the human form, including ivory carvings. For more than a century the hostility of the Moslem and Christian iconoclasts toward all graven images, especially of holy personages, periodically exploded in a fury that destroyed much representational sculpture in ivory. Although the demolishment of religious art spread to neighboring regions, some were saved by fugitives who carried small items westward to France, Germany, and other countries in Western Europe. Also, many ivory sculptors and craftsmen went into exile in the West.[18]

From the time that Charlemagne, the Frankish king, was crowned by the pope in A.D. 800 as Holy Roman Emperor (so-named because of Charlemagne's plan to revive the fallen Western Roman Empire), Christianity took a firm hold of Western Europe, and ivory was used increasingly for decorating ecclesiastical furniture, reliquaries, and book covers. Religious accessories, usually made entirely of ivory, included pyxes for consecrated wafers, retables (ornamental screens and alters), holy-water vessels, episcopal combs, and pastoral staffs.

In Russia, by the tenth century, an ivory carving industry had been started by Novgorod settlers who had colonized Pomorye. The twelfth century saw ivory carving established at Archangel. The supply of material was plentiful; enormous deposits of mammoth tusks were available in the eastern regions of Siberia. There were carvings of figurines, rings, combs, knife handles, perforated boxes, and many other articles. The imaginative genius of Russian carvers is well shown in their fine chess pieces. Some time after Archangel and Kholmogory had become the main ivory carving centers, prominence shifted to St. Petersburg (Leningrad) and Moscow.[19]

The eleventh and twelfth centuries saw great ivory craft centers arising in Germany, France, Belgium, and England. A freer spirit in art portrayal manifested itself in ivory art too. The representations became livelier and more expressive, with an occasional flair for the fantastic. Previous emphasis upon liturgical use of ivory began to give way to production of secular ivories. For example, England was soon furnishing combs, brooches, mirror cases, jewelry cases, chessmen, hunting horns, knife handles, chalices, and ivory-inlaid musical instruments. Paris became the major European center for secular art in ivory. By the end of the fourteenth century, the Italian workshop of the Embriachi family had achieved great fame, especially for luxurious combinations of ivory with precious metals. Christian devotional ivories became increasingly humanized in form and expression. The ivory Madonna sometimes had a slight smile, and for the first time she was occasionally depicted offering the exposed breast to the Christ child.[20]

Then, as religious themes yielded to secular representations, various themes supplied the subjects for portrayal on secular ivories—such as "love achieved," "love delayed," and "love lost"—typically on mirror cases and on small boxes for personal articles. Fifteenth-century ivories often showed ladies and gentlemen engaged in such diversions as hunting, riding, dancing, flirting, and playing checkers or chess.

By 1500, Portuguese navigator-adventurers had established forts on the West African coast from which they traded useful articles in return for ivory tusks and black slaves. Soon they set up similar bases on the East African coast.

The sixteenth century brought the use of mechanical aids in ivory craft work, especially the machine lathe for turning the cylindrical shapes used as complete single-piece ivory articles, or as parts of larger, jointed articles. There was also a strong demand for thinly sawed pieces that were used as inlays in furniture, weapons, and other wooden products. Significantly, inspired forms of Christian religious carvings were being phased out.

In Germany, Nuremberg's Zick family of artist-craftsmen specialized in ivory curiosities throughout their two-hundred-year succession. From the mid-1500s to the mid-1700s, more than a dozen family members of five successive generations spread the fame of those carvers and turners. The most illustrious line went from Jacob (died 1589), to Peter (1571–1629), to Lorenz (1594–1666), to Stephan (1639–1715). In Prague, Peter had taught Kaiser Rudolph II the art of ivory carving. His son Lorenz taught ivory carving to Kaiser Ferdinand III at Vienna. Lorenz's ingenuity is indicated by his so-called counterfeit boxes, carved one within another from a single block of ivory, similar to Chinese puzzle balls. His son Stephan matched his father's ingenuity by carving at least one set of trinity rings—a finger ring of triple spiral-like bands, inseparably intertwined, carved from a single small piece of ivory; one set is now at the British Museum in London.[21] Also impressive were his carvings of ivory eyes and ears which could be taken apart to reveal the mechanisms of seeing and hearing (see Anatomical Models, p. 111).

There were also German centers for ivory carving at Dresden and Regensburg; in the latter town lived the Teuber family, which produced three generations of famous carvers. In seventeenth- and eighteenth-century France, Dieppe was thought to be the best carving center in all of Europe. Featured were vases, perfume flasks, fans, jewelry, snuff graters, navettes (spindles for silk thread), and many other fine ivory articles. The following century brought a continued demand for Dieppe's exquisite ivory fans; also sought after were its mirror frames, parasols with ivory handles, paperweights, letter openers, chess sets, clock cases, and hunting knives with ivory handles. New places in Germany attracted attention as ivory carving centers. Especially distinguished were the ivory carvers of Erbach, a small village in Odenwald. Its fame began in the 1840s and soon led to prizes at foreign exhibitions; especially notable was its ivory jewelry, combined with precious metals. The master ivory carver was Otto Glenz (1865–1948).[22]

In England, some ivory carvers specialized in miniatures. Stephany and Dresch were appointed as Sculptors in Miniature to their Majesties King George III and Queen Charlotte. Later, England's artists in this unique specialty established the Royal Society of Miniature Artists. Charles Holtzapffel used machine-turning almost exclusively for his craft work with ivory and other art materials. In 1843, he published a six-volume treatise on mechanical craftsmanship.[23] Pushing machine craft to its limit for uncreative production, Benjamin Cheverton invented a "reproductive sculpturing" machine in 1828 (see p. 46). An improved model of this machine was displayed at London's Great Exhibition. (That exhibition led to the establishing of London's well-known Victoria and Albert Museum, where superlative ivory and other art works can be seen, including many of the exhibits of 1851.) It is now thought that machine-crafting of ivories—using electrically powered saws, drills, and lathes—lowered the status of all ivory craft for many years. In 1901, an Englishman created the longest continuous panel in which ivory plays a distinctive role. Frank L. Jenkins made an 80-foot-long (24 m.) frieze of bronze, including small sections of ivory (for the visible flesh parts), and mother-of-pearl, gemstones, silver, and gold. It is at Lloyd's Registry of Shipping in London, where it illustrates the history of shipping.[24]

Artistic ivory carving has continued into this century. In Italy, for example, Genoa's Borelli family ran a famous ivory workshop. Antonio Borelli is considered to be perhaps the greatest twentieth-century Italian carver.[25]

IVORY TRADE AND CARVING IN AFRICA

For many centuries now, Africa has been the major supplier of elephant ivory. The West African coast country that bears this substance's name—the Ivory Coast—was a French trading center in ivory for shipment northward across the Atlantic Ocean to Europe. Zanzibar Island, off the east coast, became the main ivory trade center for shipment across the Indian Ocean.

The pursuit and killing of African elephants became a story of adventure, excitement, and occasional death for the hunters. The transporting of equipment, supplies, and heavy tusks—conveyed by enslaved natives in later times—became a shameful story of grief, misery, and frequent death. The situation of the elephant was deplorable—slow death, or escape in a wounded and bleeding condition. Some writers have presented full accounts of this tragic situation. One is in the book by Derek Wilson and Peter Ayerst;[26] another excellent source is Ernest Moore's account of what he vividly refers to as the "scourge of Africa."[27] Moore was the Director of Ivory House in Zanzibar.

The Rennaissance in Europe stimulated an intense artistic demand for ivory; explorers and traders probed deeper into the Dark Continent and its elephant territories. Then, by the middle of the seventeenth century, the supply of ivory to the West was almost completely exhausted. In the fifteenth century, the Portuguese had started to buy slaves from African chieftans for sale to Europe and then to North and South America. In the seventeenth century, the British, Dutch, and French entered the slave trade. As the siege against elephant populations continued, the period from the 1700s to the mid-1800s saw an increase in colonial exploitation of Africa. Armed now with guns, the traders more safely ventured even deeper inland from the coasts toward the heart of Africa.

The major Eastern onslaught was started by the Sultan of Oman, who later established a trading center at Zanzibar. He invited the Western nations to trade with him for ivory tusks and enslaved blacks. Back in the elephant territories, Arab traders used products from the outside world for bartering, such as cloth, garments, beads, spices, and soap. At first they paid the natives with such things for the ivory tusks and for their services in carrying the tusks to the east coast for shipment abroad.

The second half of the nineteenth century became the climax of a bloody period of contempt for both animal and human life in the lust for ivory tusks. When the European and American traders began to offer the Arabs guns and gunpowder in exchange for tusks, the subsequent practices of many Arab ivory traders became inhumane. The worst among them no longer thought it necessary or expedient to actually pay for either the tusks or the services of the carriers. They simply combined ivory commerce with the already established African slave trade. They attacked and burned down any village that had stores of ivory tusks, and enslaved enough of the natives to carry the tusks to the east coast, where these victims were then sold into further slavery. Attached to each other by rods or chains from neck to neck, as many as half would die along the way, weakened by hunger, exhaustion, illness, and the struggle to free themselves. A half-century reign of terror gripped much of Africa for "white and black gold" (white ivory and black slaves).

The most widely known and perhaps the fiercest Arab trader, whom the natives called Tippoo Tib (various spellings) in imitation of the sound of guns fired on his human victims, was born in 1837 as Hamed bin Mohammed. He became the unofficial ruler of a virtual sultanate in Central Africa. If a village refused to barter their tusks cheaply enough, or failed at least to offer a substantial tribute contribution from their store of tusks, his armed legions burned down the village, took all the tusks, and enslaved the villagers as carriers. Often wives and children accompanied the carriers and suffered the same fate. Tippoo Tib is credited with clearing numerous African pathways along the trade routes. More dramatically, he helped European and American explorers—Livingstone, Stanley in his later charting of the Congo River, and Cameron and Wissman in their separate crossings of Central Africa. In retirement he was interviewed for the story of his life. His rambling and boastful account was later revised as a biography.[28] When he died in 1905, a wealthy man, his final hope had been left unfulfilled—to visit Europe and also the holy city of Mecca.

Meanwhile, there was a similar, though less significant, situation to the north, in the Sudan. Situated south of Egypt, it was under Egypt's control at that time. From 1840 to 1890, adventurers poured into southern Sudan and brought back to Cairo an annual supply of more than 40,000 pounds (18,000 kg.) of ivory tusks. Some were bought by barter, others taken in raids upon native villages, where the natives were sometimes captured as well and brought back north. But after British troops occupied Egypt in the 1880s, the combined ivory-slave trade began to decline rapidly. Also, slavery was becoming outlawed almost everywhere in Africa, so trading in ivory became separate.

Although Africa continues to supply most of the world's ivory, it is difficult to determine when ivory carving first started there; few tribes are able to trace their history back more than a few centuries. By the 1500s, Portuguese traders were paying African carvers to fashion particular ivories for distribution in Europe; these included spoon and fork handles, salt cellars, hunting horns, and various art objects. One art curio resembles the head of Janus, presenting two carved faces, one at the front and one at the back; sometimes the faces were of the opposite sex.[29] To expedite the carving of ivory products specifically desired in Europe, the traders brought hired native carvers to Portugal.

In the Ivory Coast and Congo regions, particularly in Benin, tusks were elaborately carved in a highly expressive style. In the Congo in the nineteenth century, a favorite carving consisted of almost an entire tusk that was decorated with a procession of carved figures advancing upward toward the end point.

IVORY IN NORTH AMERICA

For more than a thousand years, Eskimos have been using mammoth tusks, as well as fresh tusks from walrus, for carving such utilitarian products as hunting implements, tools, household utensils, boat fittings, sled runners, carrying handles, combs, and small containers. Less utilitarian articles included toys, charms, and ornaments. Artistic carvings were mainly of human and animal figures. Later, other ivory articles were added in great numbers for the commercial market. When modernization came to the Alaskan Eskimo area, ivory as a major industry was abandoned in favor of more profitable employment. The 1950s, however, brought a renewed interest in the purchase of Alaskan carvings, and soon the limited number of remaining ivory carvers became insufficient to meet the increased demand.

When walrus became scarce in the Arctic, Eskimos had to import elephant tusks. Then, ironically, many buyers refused carvings in the "substitute" substance; they wanted walrus ivory because walrus was indigenous to that part of the world. Even more ironic is the fact that the particular ivory products they wanted were often things not associated with traditional Eskimo life—gavels, letter openers, napkin rings, and so on. Unfortunately for the Eskimo carvers, serious competition developed in neighboring regions. Faked Alaskan Eskimo carvings began to appear in western Canada, and then in Washington and Oregon. Not only were many made by non-native carvers, but some were manufactured on duplicating machines (see p. 46).[30] Also, when it became evident that the tourist trade placed greater emphasis on old carvings, some Alaskan carvers quickly adapted to the demand: they "aged" new ivory carvings by boiling them in tea, coffee, and other liquids.[31]

By the end of the nineteenth century, most ivory products in the United States were utilitarian in nature, consisting mainly of piano keys, billiard balls, knife handles, and combs. Nonutilitarian ivory carving—art sculpture—had become a thing of the past. The major American experience began and ended entirely within that century, in two periods of ivory carving. The shorter period (1812–1815) concerned American prisoners of war in British camps during and after the British-American war of 1812, where they met Frenchmen captured by Britain on the seas between 1793 and 1815. Americans saw French carvers fashioning small models of ships from ivory. Art dealers and visitors to the prison camps bought the carvings; in turn, the carvers purchased ivory blocks, bones, and tools brought in by the dealers. Some Americans, without previous experience, but who apparently had the latent talent, took up ivory carving, especially of the popular ship models. Later, a few admitted quite frankly that their only purpose had been to make enough money to bribe a British guard in the hope of effecting an escape.[32]

However, for many years American whaler-seamen continued to use sperm-whale teeth for scrimshaw work (see Scrimshaw, p. 87). These carvers sold their scrimshawed whale teeth to shipmates who did not carve and to dealers in marine curio shops at ports of call.

At the end of this historical review, we may reflect upon the present location of millions of ivories from the past. Before the twentieth century, a great number of homes had at least one ivory article. The improved economic conditions of the 1700s meant that the middle classes could, for the first time, afford small ivory products. Eventually, many of the less valued ivory things were lost, or discarded when they were broken or irremediably discolored.

An ivory collector's demise, or even his severely reduced financial state, often led to the dispersal of an ivory collection that required half a lifetime to accumulate. Moments of intense feeling came when a magnificent collection of ivory carvings could find no buyer willing to purchase the entire collection. For example, the great Spitzer Collection in Paris had 175 of the finest kinds of ivory carvings, ranging from ancient Roman to modern works. During the years before the sale, thousands of invited guests had rejoiced in the opportunity of seeing such an artistic feast. Then the auction sale in 1895 scattered to museums and private collectors what was once one of the world's great ivory collections.[33] There is some consolation, however, in the fact that the ultimate dispersal of a collection does enable a wider audience to see and enjoy at least some of those art works.

II
SOURCES OF IVORY

The term "ivory" refers to a modified kind of dentin in certain types of teeth, usually one pair of tusks that extend outside the mouth and are adapted for use in obtaining food; they are also used sometimes for attack or defense. The following discussion begins with elephant ivory, since this is the most common source. As will be noted, substantial quantities of ivory are also obtained from the teeth of two non-tusked species: the hippopotamus and the sperm whale.

ELEPHANT

Elephants have long been hunted by man, not only for tusks, but also for meat, oil, and hide. Their tusks are the two upper incisor teeth, that is, the front teeth, not the canines as in other tusked species. The tusks are deeply embedded in the upper jaw and the skull, thus providing a strong anchor needed for heavy lifting, as well as meeting the primary requirement of supporting the heavy tusks themselves. The length of tusks varies according to the species and sex of the elephant: tusks from African elephants are generally larger than those from the Asiatic species, and males usually have longer, heavier tusks than the females of their species. Among the members of one sex of a species, there are also substantial individual differences.

In the male Indian elephant, the tusks are seldom longer than 8 feet (2.5 m.) and usually no heavier than 90 pounds (41 kg.) each. The female, being smaller, has smaller tusks, which sometimes hardly protrude from the mouth at all. In nearby Sri Lanka, even male elephants are often without noticeable tusks.

The main source of ivory has been the African elephant. At least in the past, some males had tusks over 10 feet (3 m.) long. The jaw-embedded part accounted for as much as 30 inches (76 cm.) of this length. Such tusks may weigh 150 pounds (68 kg.) or more and may be over 8 inches (20 cm.) wide at the base. Although large tusks were prized, small ones have not been without value; in England's ivory markets, they were designated as "scrivelloes," and were usually defined as being less than 18 pounds (8 kg.) weight.

Particularly in the African species, the tusks are used for uprooting vegetation, tearing edible bark off trees, lifting or pushing away obstructions in the elephant's path, and for fighting if necessary. The female is smaller and has shorter and straighter tusks; billiard balls were usually cut from these and, as a consequence, the ivory they furnished was termed "ball ivory."

For most, if not all, of an elephant's life, the growing tusks are built up in thin layers of conical accretions on the walls of the tusk's hollow interior. Because living ivory is elastic, the tusk does not crack as the outer layers gradually expand to accommodate the continuing inner enlargement. It has been reported that ancient disintegrating ivory has sometimes come apart in conical form.[1] Of the visible tusk that extends outward from the jaw, the hollow section is almost half that extent; it contains the nutritive pulp-tissue and the nerve. This hollow shaft has been referred to as "bamboo ivory" because of its form. Enamel coating is seldom if ever seen on elephant tusks. They do have a brownish outer layer, or bark, which has sometimes been retained as the unseen bottom surface of an ivory carving.

As a modification of ordinary dentin, ivory consists mainly of calcium phosphate. The density of elephant ivory relative to that of water (i.e., its specific gravity), is usually reported to be between 1.83 and 1.93. Bone is less dense. Depending on the animal's geographic region and the angle at which the tusk is cut, some ivory shows a grainy effect. A surface cut lengthwise may show wavy lines, or even a conic or parabolic effect. Some cross-sectional surfaces of cut ivory show a mass of tiny adjacent diamond-shaped figures, patterned in arcs that present an appealing criss-cross or checkerboard design (Fig. 6). As will be noted later, this criss-cross pattern is one of the very few simple and fairly reliable indications of genuine ivory (see p. 60). Unfortunately, most ivory lacks this grainy effect. Grain in some ivories may become more visible if the ivory piece is tilted at various angles toward a good source of light.

Because of its oily element, ivory yields a polished appearance when rubbed with a soft cloth. After a long period of time, the ivory may dry out; some collectors use a transparent polish for adding a surface gloss. Most ivory darkens with age, becoming yellow or brown and sometimes, after many centuries, resembling any of various kinds of wood. The surprising flexibility of fresh, or "green," ivory was evident in riding whips; thin strips of fresh ivory are easily bent back and forth. As a poor conductor of heat, ivory was fittingly used for the handles of teapots. As a poor conductor of electricity, ivory discs and small cylinders were once used as electrical insulators (see Fig. 46).

Although the term "elephant ivory" refers to the two tusks, the molars (grinders) teeth have been used as a substitute. But because they do not have a homogenous material throughout, they have been used for plain articles only, such as small boxes, knife handles, and cheap ornaments.

Game laws now restrict the hunting of elephants. Animal reserves have been designated in which hunting is entirely forbidden. Other areas require a license that permit a limited number of animals to be taken. Poachers, however, continue their illegal hunting, aided by natives whose share of the catch often consists of the elephant meat and all else except the ivory tusks. Ironically, and illustrative of how human intervention creates its own imbalance, the conservation movements have led to such rapid increases of elephant populations that some herds have begun to consume all the available foodstuffs. Large regions of forest and bush have become defoliated by the elephants; in the worst areas, hunters had to be invited in to thin out their numbers.

FOSSIL: MASTODON AND MAMMOTH

The elephant's predecessors were the mastodon and then the mammoth, which once ranged widely on much of this planet. The largest reserves of those extinct species have been found in Siberia, buried in frozen land masses. Fossil tusks dug up from frozen Arctic shores were sometimes called "beach ivory." These animals were larger than the modern elephants, and their tusks were longer, heavier, and severely curved, mostly upward. Although the tusks of extinct species have often become altered in composition and appearance, the term "fossil ivory" is actually incorrect; most of those tusks have not yet undergone the substantial physical conversion (i.e., mineralization) implied. Such a conversion would render the substance unsatisfactory for use as what we know and appreciate as ivory. Some ancient tusks, such as those discovered in Egyptian sand deposits, were in remarkably good condition and appearance.[2]

Interestingly, carvers have intentionally used the discolored ivory of mastodon tusk, carving small articles of costume jewelry such as brooches and pendants of varied colors.[3] Attractive "gemstones" have been cut from fossil ivory which had mineralized into a colorful and very hard condition. From the frozen north, the substance was sometimes a brilliant blue color and was used as a substitute for turquoise.[4]

WALRUS

The walrus (from the Scandinavian for "whale horse") was hunted by Eskimos and other natives of the far north for its meat, oil, hide, and tusks. In the male, the two slightly curved canine tusks, pointing downward from the upper jaw, have an average length of about 20 inches (50 cm.) and an average weight of about 6 pounds (3 kg.) each. The jaw-embedded section has a hollow interior and adds about a further 10 inches (25 cm.) to the overall length. The female tusks are shorter, more slender, and therefore lighter in weight. In cross section, the tusk is oval shaped and does not show the intersecting lines seen in some types of elephant ivory.

Normally the walrus uses its tusks only for digging out its food on the ocean shore—shrimps, clamlike animals, and edible plants. The strong pair of tusks also aid in climbing up slippery rocks and in crossing icy ledges. It may also use its tusks to defend itself from an attack on land by a polar bear.

Ivory carving from walrus tusk was prominent in northern Europe, especially in Scandinavian countries. The earliest known name of a carver of walrus ivory is "Margret, the priest's wife," in Iceland. A personal communication to the present writer from the National Museum of Iceland tells that she is mentioned in a saga written by a bishop who died in 1211; she had carved an admired ivory crosier that he presented to the Archbishop of Norway. Medieval Scandinavian warriors thought that ivory handles on knives, swords, and javelins, as well as ivory rings on fingers, could ward off the ailment of cramps; the substance was probably from walrus tusk. Similar was the Near Eastern Moslem belief that walrus-ivory hilts on daggers had the magical property of stanching the flow of blood and inducing faster healing of wounds if the ivory were applied to the wound. It was said that if food had been poisoned (a favored way of disposing of one's enemies), walrus ivory would show this by starting to "sweat" as soon as the ivory was brought to deadly food.[5]

English buyers in the sixteenth century were paying twice as much for walrus ivory as for elephant ivory. In India, the Emperor of Agra was so delighted with a Persian dagger that had a hilt of walrus ivory, he dispatched his traders to Persia with instructions to buy walrus tusks at any price. Perhaps this enthusiasm was related to the spreading belief that walrus ivory was an antidote to poison and could also reduce the swelling from infections. A Turkish treatise of 1512 tells that poison has no effect upon a person who carries a walrus tusk with him.[6]

In seventeenth-century China, walrus-ivory chopsticks sold for six times the price of elephant-ivory chopsticks, because of the reputation for detecting poisoned food. An unusual walrus-ivory article was the Chinese abstinence plaque, worn as a pendant that indicated to others that the wearer was making a religious fast and should therefore not be offered certain luxury foods.[7] At one time, some Chinese carvers dyed walrus-ivory ornaments a green color in order to sell them as jade.

Diminishing numbers of walrus after centuries of hunter forays led to a great decline in the supply of walrus ivory after 1700. Alaskan Eskimos had a steady if limited supply, and continued to use it for making their tools, as well as ornaments and articles for the tourist and export trade. Ivory was essential for tools there, because wood was scarce, and metal came late to that part of the world.

NARWHAL

The narwhal (or narwhale) was once known as "the sea unicorn" because of its single outgrowing tusk. It has been hunted in the cold Arctic seas, and sometimes down to a latitude of 60 degrees north, for its long straight tusk and as a source of food and oil. Eskimos also used its leathery hide, and its intestines, when emptied and dried out, served as lengths of cord. It is a small type of whale, usually 12 to 16 feet (3.5–5 m.) long. The origin of its name is indeed strange. In Old Norse, *narwhal* meant "corpse whale," so-named because its white color made it look like a floating corpse.

At an early age, the narwhal loses almost all its teeth. The male retains the upper left canine tooth, which then grows forward as a long, straight, spirally twisted tusk. In some instances, the right upper canine tooth shows a slight growth. The single prominent tusk averages 6 feet (1.8 m.) in length (about half the animal's body length), weighing about 9 pounds (4 kg.).

Speculation about the function of the tusk includes two major possibilities: as an offensive weapon in fighting other males for possession of a female narwhal; as a defensive weapon against predators of the sea. It may also be used to crack overhead ice sheets when the male comes up to breathe air; of course, the tuskless females must depend on the males for this operation.

The color of the ivory is a pale white. The cross-sectionally cut surface sometimes shows concentric rings and radiating, crossed lines.[8] The ivory is of fine quality and does not appear to yellow with age. Because the tusk is slender and is hollow for more than half its length, it has been used primarily for small carvings such as napkin rings and Japanese netsuke. Often, the attractive tusk was kept whole as a natural curio, displayed in a case or mounted on a wall. One is at Chicago's Field Museum of Natural History.

Vikings of the eighth to twelfth centuries placed great value on narwhal tusks, partly because of the difficulty and danger in catching the narwhal (which has its own sword), and partly because they admired the attractive appearance of the tusk and the texture of the ivory. Vikings decorated the prows of their war vessels with entire narwhal tusks, and they carved whole tusks into swords, staffs, and scepters. Ancient Chinese used ground narwhal ivory in solution as a medicinal cure.

In the fourteenth, fifteenth, and sixteenth centuries, narwhal use and cost grew enormously because of a belief, similar to that about walrus, that it could reveal the presence of poison by becoming moist. Being both quiet and cheap, poison continued to be a favored method for disposing of rivals and enemies. A narwhal-ivory drinking cup was thus a choice possession. According to a 1636 document from Prague, a narwhal tusk was tested and was found to be unable to detect poison; so the gem dealers declared that it could not be genuine narwhal tusk. The great coronation chair of the kings of Denmark (first used in 1671 and still seen there) is profusely decorated with ivory inlays from narwhal tusk.[9]

As in the case of elephants and walrus, species-preservation laws and international agreements have recently restricted the killing and commerce in the by-products of several species of whales, including this unusual kind.

HIPPOPOTAMUS

This huge African animal is mainly aquatic; indeed, its name means "river horse." It is the second heaviest water animal, next to the larger species of whales. As a land animal for part of its day, it is also the second heaviest, next to elephants. The hippo, now a government-protected species in most African countries, was hunted for its meat, oil, hide, and teeth. Although all of the front teeth are of ivory-like substance, our major interest is in two of the six front teeth in the lower jaw. Two strong lower canines, used for uprooting rank grass (congested grass growth), are deeply embedded in the bony sockets of the jaw. The hippo also sometimes uses these elongated canines to defend itself against an attack by crocodiles. They are entirely contained within the massive mouth, and therefore cannot be seen unless the great jaws are opened wide. Because even these long teeth do not extend beyond the mouth, some writers do not consider them to be tusks.

These two canines are more curved than elephant or walrus tusks, forming almost half a circle. The cross-sectionally cut surface is slightly triangular. The outer layer of protective enamel is extremely hard and was very difficult for ivory workers to remove; steel tools often struck off sparks of fire when forcefully applied to this hard surface. The difficulty in removing the enamel coating made these teeth a less-favored substance for ivory workers. By the 1800s, however, the task became much easier when the rotary grindstone became available.

The ivory is pure white and quite hard; it hardly yellows with age. Being so dense, it resists stains. It was thus favored for making artificial teeth and sometimes for small ruler-like scales on scientific instruments. In solid form, the curved end was often used for handles on umbrellas, canes, and doors.

BOAR AND WARTHOG

Wild boars are now found mainly in North Africa and in central and southwestern Asia. Two short but powerful and sharp canine tusks are in the lower jaw; two larger ones—sometimes 10 inches (25 cm.) long in the male—are in the upper jaw. One of the last remaining species of wild pigs that have four tusks, these animals have the type of tusks—all projecting upward—that are effective for both uprooting and fighting. The small size of the tusks has kept their use limited to such things as handles and ornaments. This is also true for the tusks of smaller wild pigs such as the warthog and the babirusa.

SPERM-WHALE TEETH

Sperm whales are found in all oceans, living and traveling in groups. Since they must periodically rise to the surface to breathe, they have been easy prey for hunters. A thick layer of fatty blubber insulates them from the cold water and was the main substance for which they were hunted. Sperm whales are the largest of the toothed whales. Lack of substantial teeth in the upper jaw is not a disadvantage, because the lower jaw teeth are used mainly for retaining the trapped fish and squids, not for chewing; the food is swallowed without being masticated. This huge sea animal has also been hunted for its superior oil, spermaceti (for cosmetics and ointments), meat, whalebone, and the several dozen large teeth in the lower jaw. These teeth are a form of ivory and may be up to 8 inches (20 cm.) long and 3 inches (7.5 cm.) wide, weighing about $2\frac{1}{2}$ pounds (1 kg.) each.

Because the color of the inner substance is somewhat dark, the teeth have not been valued highly as a source of ivory. Furthermore, because the first half is hollow, craftsmen preferred to use the whole tooth, instead of parts of it. They mainly established an art form of etching the teeth, rather than carving. During the nineteenth century, English and American whaler-seamen whiled away a part of their leisure time by etching scrimshaw work on whole teeth, as well as on whalebone and on seashells. Also, a variety of small useful articles would sometimes be carved from the teeth, although whalebone was a more popular material in this respect.

Whale teeth and bone gradually became less available as more and more countries agreed to prohibit, or at least limit, the killing of certain kinds of whales. Some countries that do not have a tradition of hunting whales agreed to prohibit the commerce in whale by-products.

III
IVORY SUBSTITUTES

Ivory became the preferred substance for many kinds of artistic and utilitarian carvings, but the supply was seldom sufficient. Thus, it was inevitable that substitutes would be used, most of which closely resembled ivory. The first substitute discussed below was much more plentiful than ivory and apparently predated its use, especially for utilitarian products.

BONE

Because almost all the animal species encountered by man had a body framework of bone—strong, but light in weight—such an easily available substance has long been used for practical articles, and sometimes for artistic carvings. Even though bone items resemble ivory, examination can often detect the difference. Bone lacks the grainy appearance of some ivory, as well as the mellow and creamy color. The bone cavity has a spongy substance, which is noticeable on any carving that exposes the interior. The cross-sectional surface has tiny holes that appear as dark spots or as a pitted surface. The longitudinal surface often shows short dark lines—the canals that pipe the vital fluids in living bone. Commercially, oil or wax was sometimes rubbed into the surface in order to cover such spots and marks, thus increasing the difficulty of distinguishing bone from ivory.

When large sizes were needed (as for single-piece canes, measuring sticks, and blackboard pointers), bone workers used large pieces of whalebone from the huge jawbone of the whale. The craftsmen often embellished their utilitarian products in an artistic manner. Some scrimshanders decorated the entire jawbone of a porpoise as a kind of artistic curio.[1]

Among the numerous practical uses of bone were buttons, shoehorns, clothespins, boxes, handles, door knobs, dominoes, and dice ("bones" in gaming slang). Bone dust, the residue from sawing and carving, was sometimes used in a hardened compound for making small articles, similar to one use of ivory dust. Almost no whalebone can be obtained now; even in countries that still engage in whale killing for commercial use, all the bones are usually pulverized into a constituent of chicken feed.

A substance similar to whalebone comes from baleen whales. From the upper jaw, firm but flexible sheets of baleen hang down inside the mouth; coarse hairs are attached to these sheets. The function is that of a strainer, so that this type of whale can take in a mouthful of small marine animals in the water and then expel only the water, while the baleen vertical strips and attached hairs retain the food. Once incorrectly termed "whalebone," baleen was used for corset stays, hoopskirt hem-frames, perimeter frames for hats, and many other articles in which the dark color of baleen did not matter. In later times, baleen was used for canes, shoehorns, penholders, letter openers, paper creasers, trunk framework, carriage wheels, and other useful products. It was sometimes even used for simple ornamental inlays.[2]

HORN

Horns on the top of an animal's head are usually used for fighting. From ancient times, mankind has used the horn for making daggers, sword hilts, harpoons, powder horns, handles, spoons, drinking cups, small containers, buttons, drawer knobs, and other items. Some of those horn products became possible after craftsmen discovered that moist, heated horn becomes malleable under pressure.

When a hole was made at the end point, horns were used as one of the oldest of sound instruments, and to which they gave their name. Small pieces of stag horn were once used for decorative inlays on the wooden surfaces of weapons such as crossbows and, at a later time, on firearms, similar to the use of ivory inlays on weapon surfaces.

The several varieties of African and Asian rhinoceros were hunted for their meat, hide, and the single horn of the Asiatic rhino or double horn of the African rhino. In addition to use in fighting, the horn serves as a tool for uprooting bushes and shrubbery in the rhino's search for leaves to feed upon.

Rhino horn has been used for carving small artistic pieces, including Japanese netsuke. Because the hunting of this animal for mere sport had left some types almost extinct, many countries placed severe restrictions on rhino hunting and on commerce in its by-products.

In closing this section, it should be noted that horn is an outgrowth of the skin and does not contain dentin or bone. Because its appearance could hardly be mistaken for ivory, horn should be considered an unpretentious alternative to ivory, rather than a mere substitute.

HORNBILL

Of the many varieties of hornbill bird, only the helmeted hornbill furnishes an ivory-like substance. At the front of the head above the beak, this rather ugly bird has a casque as dense as elephant ivory. The maximum size of the helmet is 2 inches (5 cm.) long, 1½ inches (3.8 cm.) high, and ½ inch (1.3 cm.) wide. Living in the highest trees in the Malay Peninsula and other parts of the East Indies, the bird is seldom seen from the ground, despite its extensive 5-foot (1.5 m.) length from the tip of its large beak to the end of its tail.

Because of the substance's warm yellow or orange color, it became a prized material for miniature carvings as early as the fourteenth century. Besides the East Indies, China too used it for small carvings. Later, Japan bought these casques particularly for carving the dainty sliding bead *(ojime)* that loosened and retightened the two parallel cords that made it possible to open and close a small portable container *(inro)*.

By the beginning of the twentieth century, the bird was becoming scarce, victim to the value of its casque. At that time a casque was selling for about $20. In 1960, when this species had become almost extinct, hornbill snuff bottles were selling for about $300, and sometimes twice that price.[3] By comparison, attractive elephant-ivory snuff bottles were then being sold for about $50.

Besides brooches, a favorite Chinese carving was the two-piece rectangular buckle. Shuyler Cammann's article on hornbill illustrates one such item from New York's American Museum of Natural History and another one from Chicago's Field Museum of Natural History. Also shown from the Chicago museum is an archer's thumb-ring used for drawing back the taut bowstring. A Chinese snuff bottle is shown from the Seattle Art Museum. The casque was sometimes used for an artistic carving; Cammann shows one from the University of Pennsylvania Museum, and two from the Chicago museum.[4] In Borneo, earrings and ornaments were made from this substance. Finger rings were carved in Malaya, where it was believed that a hornbill ring would detect poisonous foods by changing in color when adjacent to them. Because some products are longer than the maximum length of the natural casque, Cammann conjectured that the Chinese craftsmen of such products had flattened the casque.[5]

VEGETABLE IVORY

Except for the moderate supply of tusks from land animals discovered dead from natural causes, supplying ivory requires that the animals be hunted, slain, and, in the case of sea animals, retrieved. It is not therefore surprising that these difficulties, and consequent shortages in supply, led to a search for substitutes. When earlier substitutes became insufficient or came to be considered unsatisfactory, attention turned to so-called vegetable ivory. Soon thereafter chemically synthesized substitutes became available.

The principal source of vegetable ivory was the inner seed of the nut from the ivory-palm tree, *Phytelephas macrocarpa* (Greek: *phyt*, plant; *elephas*, elephant ivory; *macro*, large; *carpa*, fruit). This stunted palm grows mainly in the damp western climates of South America—in Ecuador, Peru, Colombia, and Brazil—and is known by various local names, including *tagua*, *yarina*, and *corozo* (or *corosos*). The seeds are the shape and size of hens' eggs, with an average weight of about $1\frac{1}{2}$ oz. (45 g.). They have little taste and were not therefore marketed as an edible nut. Before use, the seeds must be dried out. They become very hard, though not as hard as ivory. The substance is then smooth, takes a good polish, and easily absorbs dyes for decorative purposes. However, it does have a dull appearance in comparison to ivory. Its specific gravity is reported to be about 1.4, somewhat lower than that of ivory. It consists almost entirely of a form of plant cellulose.

The ivory-palm nut has been used mainly for buttons, but also for thimbles, dice, checkers and chess pieces, poker-game chips, knobs on the handles of umbrellas and walking sticks, bottle stoppers, needle cases, ornaments, figurines, and artificial flowers. Sawdust waste from the nut was used as a polishing agent and as a cattle-feed supplement. A lengthy article has been published about this commodity.[6] Despite its great value in the economy of Ecuador and several other South American countries, very little concerning this nut had been published in South America until the Director of Ecuador's Institute of Natural Sciences wrote a treatise about it in 1944.[7]

Natives from districts growing ivory-palm nuts were reported to be making buttons from this substance as early as the 1840s. During the first third of the twentieth century, millions of pounds of the nuts were imported annually into the United States and Europe, mainly for button manufacture. Figure 1(a) shows examples of carvings in ivory-palm nut.

The so-called gingerbread nut from the doum palm (also "dum" or "doom"), growing in parts of northeastern Africa and some places in the Middle East, served similarly as a source of vegetable ivory. Demand for a cheaper button material brought it into commerce as a substitute for the ivory-palm nut. Doum-palm nuts are yellowish brown, irregularly shaped, and smaller and cheaper than ivory-palm nuts. However, they do not harden as well, and the material would sometimes shrivel and warp after the buttons had been made.[8]

In India and Southeast Asia, a tall slim palm tree *(Areca catechu)* yields the betel nuts that have been chewed since ancient times. This substance has been used for small articles, but is not really comparable to ivory-palm nuts or even to doum-palm nuts—the white nut-meat is distributed throughout the brown body of the fruit in a streaked form, so that the entire white and brown dried fruit has to be used in the articles. Favorite items include buttons, small knobs, ball-shaped handles for key rings, and costume jewelry. Betel nuts and a key ring are shown in Figure 1(b). The Department of Agriculture in India informed this writer in 1978 that it was not aware of any research or publications concerning the carving or utilitarian manufacture from these nuts.

Early in the twentieth century, a treatise on ivory mentioned that an "older system" for making vegetable ivory consisted of boiling a potato in sulphuric acid.[9] The present writer found this claim surprising, in view of the known corrosive nature of that strong acid. He then boiled potato pieces (both red skin and white skin) in three separate concentrations: 10, 25, and 50 percent. The pieces not only became softer, as they would in even plain boiling water, but they gradually became a purplish black, gelatinous mass. So that earlier recommendation is indeed strange.

ARTIFICIAL IVORY

Because all vegetable ivory-substitutes had limitations and disadvantages (very small size, lack of close resemblance to ivory, foreign shipping problems), chemical substitutes were soon attempted. Since World War II, vegetable ivory has largely been replaced by various synthetic compounds. One early compound consisted of caoutchouc (natural rubber), sulfur, and a white coloring, such as gypsum, white clay, or zinc oxide. Figure 1(c) shows an india-rubber button (patented in 1851) made without the white coloring that might have made it resemble ivory.

By 1863, the Phelan & Collender Company, largest manufacturer of ivory billiard balls in the United States, had become so concerned about the increasing shortage and cost of ivory that it offered a $10,000 prize to the first person who could produce a good synthetic substitute (see Billiard Balls, p. 122). Two Years later, in New York State, John Hyatt concocted a substance based on collodian—a highly flammable nitrocellulose solution—which, with an ivory color added, could be hardened into solid form. Although his invention was not deemed close enough to ivory to warrant receiving the prize, Hyatt started the Embossing Company of Albany (New York) for the manufacture of dominoes and checkers.[10] In 1867, he improved the formula, simulating ivory more closely, and he founded the Albany Billiard Ball Company. His partner brother coined the name Celluloid. Soon they were making a variety of products for which the ivory substance had become too expensive or was impracticable.

Competing manufacturers were making similar substitutes, under such trade names as Xylonite, Cellonite, and Pyralin. In later years, some synthetics would bear more appealing (and deceptive) names: French Ivory, Ivorine, Ivoroid, and so on. The new substances were stronger than ivory and more resistant to breaking or splintering. They were lighter in weight, insoluble in water and stain resistant; dirt could easily be removed with a moist cloth. The uses soon included fountain pens, soap dishes, toothbrush handles, shaving-brush handles, combs, hairbrush backs, small mirror and picture frames, small boxes, clock cases, collars, letter openers, as well as containers for liquids, lotions, and ointments. Especially popular were the complete sets of related items, such as dresser sets and desk sets.

Later improvements added the longitudinal grainy effect seen in some ivory (Fig. 8). But there was a serious hazard with Celluloid and any other synthetics that used nitrocellulose—flammability. If accidentally brought near fire, they could suddenly burn up. Newer synthetics avoided this hazard.

Although most artificial ivory was too hard to allow carving of ivory-like artistic pieces, one synthetic soon claimed success. Around 1890, Alexander and Silvius de Pont, two Swiss citizens residing in Paris, invented a compound that somewhat resembled the constituents of actual ivory. It consisted of the following, in decreasing proportions: water, caustic lime, aqueous phosphoric acid, albumen, carbonate of lime, gelatin, alumina, and magnesia. The portions, added in a specified sequence, were mixed and kneaded, and then dried out, either in shaped molds for ready-made figurines or useful articles, or in block pieces that could be carved or manufactured into desired items. The resultant substance was said to be nonflammable, insoluble, and similar to ivory in appearance.[11] We may wonder about the subsequent use and artistic acceptance of that promising ivory substitute.

IV

CUTTING AND CARVING

———•◆•———

The first cutting of animal ivory requires an efficient removal of the tusk from the dead animal's jaw. Although a saw can cut the tusk at the point where it leaves the jaw, this would omit the part of the hollow section embedded within the jaw. For large tusks, as those of elephants, the loss could be more than 20 inches (50 cm.). Even though the hollow section is of much less value, some hunters used a sharp ax to laboriously chip away the bony socket in the jaw. The hollow section has been ideal for making cylindrical lampshades and containers.

PREPARATION OF THE IVORY

As in the case of carving wood, ivory first requires seasoning (drying) to allow for natural shrinkage; otherwise the fashioned items would gradually suffer shrinkage and distortion—weight losses of up to 4 percent have been reported. Ivory, however, has often been used before such shrinkage has taken place. In northwest India, tusks were sometimes coated with melted wax to keep them fresh in storage until they were taken up for carving.[1]

Billiard balls are probably the best example of ivory articles which require perfect seasoning before the final cutting. The ball must be absolutely spherical in order for it to roll along a straight, predictable line when the ball is struck. After some initial drying out, the ivory was machine-cut into spherical form, a little larger—about $\frac{1}{16}$ inch (1.5 mm.)—than its intended later size. Then, after several more months of the contraction process under carefully controlled conditions, the ivory ball was cut down to its final size. In an era when manufacturers could rely on the long-term operation of that industry, it was not unusual for a stable enterprise to season the slightly oversize ball for as long as five years before cutting away the outer layer.

As far as elephant tusks themselves are concerned, first the brownish outer barklike layer is removed. Even this rind has been used, in the form of narrow strips, or spills, for penknife handles. But sometimes this layer was not removed and became the uncarved and unseen bottom surface of certain ivory articles.

For the cutting and carving of ivory, the tools are very diversified—saws, shears, rasps, files, drills, punches, chisels, picks, scrapers, and a variety of knives and similar cutting instruments. In recent centuries, besides those manual tools, mechanized devices have been used, making the work easier, faster, and more accurate. These include mechanical and electrical saws, drills, and lathes. Ivory is sometimes sawn in water to make it easier to cut, prevent cracking, and to cool a high-speed saw blade. Water's softening effect on the surface of ivory is shown by the fact that scissors can cut a thin sheet after it has been soaked in warm water for an hour. Some carvers keep an ivory block wrapped in wet cloths for a few days before starting to carve the surface; others tell of using cold cream emulsion instead of water.

The plan for dividing a tusk into workable sections requires analytical capacity and considerable experience in carving. Attention must be centered on the hollow and solid portions, the curved form of the tusk, the elliptical or irregular shape of the circumference of the tusk, and of course the size and shape of the articles intended to be fashioned. Because ivory's hardness introduces an additional difficulty, beginners usually learn to work on softer substances such as hardwoods or vegetable ivory.

Defects, cracks, or discolorations are sometimes discovered during the sawing or carving operations. Such faults will limit the use of that piece of ivory and will reduce the value of the finished product. Bullets have been found embedded in elephant tusks, apparently resulting from the hunter's poorly aimed shots. Holtzapffel reported in the early nineteenth century that silver and gold bullets found in tusks were probably from guns fired by rich Eastern potentates.[2]

When an ivory production required a set of matched items—such as game pieces, tableware handles, or small pieces joined together to make a statuette or a large utilitarian article—the carver preferred to use a single tusk. The reason is that it is often difficult to match pieces from different tusks so that all the pieces will be identical in surface color and texture. Even a single tusk sometimes does not provide the high degree of similarity desired.

For machine-tooled ivory work, great skill and imaginative conception were achieved by the turners in Germany. Stunning illustrations of German turning are shown in O. M. Dalton's catalog.[3] Examples of this type of work are also represented in many of the illustrations in Part Two of the present book. The most complete treatment of machine-working of ivory (and wood) was the six-volume set by Holtzapffel, cited previously. Volume 5 shows several photographs of large complex ivory productions of an artistic and architectural nature, including elaborate towers.

The working conditions in ivory factories were often primitive and hardly rewarding. In the Orient the situation was far from ideal, even at the beginning of the twentieth century. Ivory workers labored in small congested spaces under conditions and duplication requirements that precluded creative artistic expresssion. In China two daily meals were provided, plus very small wages. In earlier days, an apprentice received almost no wages at all for several years.

ATTEMPTS AT SHEET-IVORY PRODUCTION

Since early times, there has been interest in the production of flat ivory sheets wider than the maximum diameter of elephant tusks—about 8 inches (20 cm.). How did Phidias manage to cover with ivory the face and hands of his huge statues of Athena and Zeus in the fifth century B.C.? Of course, he may have used small thin strips of ivory, overlaid upon each other; or small ivory mosaic pieces may have been fitted together on the surface. The same question arises about statues made entirely of ivory mentioned in Greek and Roman writings. It is possible that the ivory face and hands of Phidias's massive statues—and, perhaps, the all-ivory statues in Greece and Rome— may have consisted of solid cemented sections of ivory throughout. We will probably never know; those statues have not survived.

A continuing demand for small flat ivory surfaces in later times is indicated by the fact that by 1700 small ivory squares and rectangles began to be used for miniature paintings, because of ivory's appealing luminous surface. Ivory Venetian snuffboxes were decorated with small portraits. Rosalba Carriera (1675–1757) firmly established the use of ivory for portrait miniatures, exploiting the warm mellow tone to convey the texture and appearance of skin. From Italy, her fame and influence spread to Germany, Scandinavia, England, France, and the United States. Small flat pieces could be easily cut from ivory tusk, thereby posing no technical problem.

Surviving book covers made of ivory, as large as 12 inches (30 cm.) square —wider than the diameter of any tusks—later led some writers to suggest that the art of softening and then flattening out an ivory piece more broadly may have been known in early times. We do know of various ancient and medieval formulas for allegedly softening ivory.[4] Modern attempts with those recommendations have failed to make ivory soft or malleable. For example, dilute muriatic (hydrochloric) acid was one of the recommendations in a fifteenth-century English manuscript concerning Italian techniques.

The present writer tried this with small ivory pieces in different concentrations of acid solution (5, 10, 25, and 50 percent). There was no softening, only the beginning of disintegration on the surface.

Early this century it was reported that in 1848 Elsner had claimed that ivory can indeed be softened and then rehardened. The ivory, it was said, has to be kept in a solution of phosphoric acid of specific gravity 1.13 until it becomes semitransparent and soft. When washed, it can be flattened or shaped into some other desired form. After drying, it becomes hard once more; subsequently, plain water will soften it again.[5] This claim has gained wide acceptance. In 1938 one writer noted that ivory tablets were being cut around the tusk and then softened in phosphoric acid, this presumably being the method by which ivory sheets as large as 30 by 30 inches (75×75 cm.) were produced for portrait painters such as Thorburn, Newton, and Ross.[6] In 1952 Thomas Penniman added an important note concerning the phosphoric acid method—that constrained reshaping is necessary.[7] Thus, a sheet can be cut around a tusk, then softened, and lastly, kept flat under pressure long enough for it to remain flat when dried out and released.

But what about Elsner's admission that the ivory will become semitransparent? The present writer tried this phosphoric acid method on a cylindrical piece of ivory, $\frac{3}{4}$ inch (20 mm.) long and $\frac{1}{4}$ inch (6 mm.) in diameter. Gradually, over several days, the piece became not only semitransparent but also amber colored because of certain chemical changes; therefore it was no longer ivory. It was soft, even flexible, but not malleable: when squeezed or bent, it simply returned to its regular form. In hours it dried out and became hard; it was still semitransparent and of amber color, but no longer of value or interest as ivory. (In water, it became soft and flexible again in one hour.) If a constrained shape was imposed upon it in its soft form and was sustained while it dried out, it would maintain its new shape when released from constraint. When placed in water again, it softened to its original cylindrical form.

Thus, this method of softening ivory would seem to require immersion in dilute phosphoric acid for a very brief period of time, terminating before the ivory's constituents change enough to produce a permanent semitransparent amber color. It is doubtful that thick pieces of ivory would become softened and still look like ivory. As a final note, it should not be thought that Elsner's recommended specific gravity (1.13) is given as absolute; undoubtedly, other concentrations have been used effectively.

Moreover, thin sheets or strips of ivory can be softened by immersion in other liquids too. Using both thick and thin samples of ivory, this writer tried several liquids recommended in later times—a mixture of vinegar and nut oil; liquid mustard; a mixture of saltpeter (potassium nitrate) and white wine; and plain water. Solid pieces $\frac{1}{4}$ inch (6 mm.) thick did not soften.

In 24 hours small flat slabs only $\frac{1}{16}$ inch (1.5 mm.) thick became more flexible in all the liquids, but did not become malleable. If held in a curved form in a vise until dried out, they retained that form when released from constraint. If later immersed in water, they softened again and gradually returned to their original flat form. Note that in the middle of the nineteenth century Holtzapffel concluded that there was no known way of softening and then restoring a solid piece of ivory.[8]

Cutting ivory sheets from around the tusk has already been mentioned. Antoine Quatremere de Quincy (1755–1849), a French archeologist, collected information about classical statues. He was especially interested in Phidias's monumental Greek statue of Zeus (renamed Jupiter by the Romans) at Olympia, with its face and hands of ivory. In a treatise published in 1815 in Paris *(Olympian Jupiter, the Art of Ancient Sculpture)*, he conjectured that the ancient carvers had probably handcut a veneer cylinder of ivory from around the outside of the tusk and split the thin cylinder into small surfaces. Then they *somehow* softened those surface sections and mounted them upon the underlying curved structures of the face and hands. Medieval book covers may have been made that way also.

MECHANICAL CUTTING METHODS

Finally, a machine was invented for cutting thin sheets of ivory from around the tusk, similar to earlier machines for cutting wood veneer from around a section of tree trunk. The Science Museum in South Kensington, London, informed the author that H. Pape received a French patent in 1826 for an ivory-veneer cutting machine, and that he displayed a 30- by 150-inch (76 × 380 cm.) ivory veneer in England in 1834. Holtzapffel wrote that Pape was also an inventive Parisian manufacturer of pianos. One of his ivory sheet cuttings—38 by 17 inches (96 × 43 cm.) in a thickness of $\frac{1}{30}$ inch (0.8 mm.), glued upon a flat board—was displayed at the Polytechnic Exhibition in London's Regent Street; he covered all the external surfaces of a piano with such ivory sheets.[9] We may wonder where that first ivory-sheet cutting machine is now? And where is that piano with a complete ivory surface? Were any other pianos made that way?

In England Cheverton patented a similar machine in 1844. In the personal communication just mentioned concerning Pape's prior invention, London's Science Museum indicated that it does not know the present location of Britain's first machine for cutting ivory veneer.

In America a race started for the invention of an ivory-sheet cutting machine. A Mr. Hite, a New York painter of miniatures, offered a generous price for ivory sheets 12 inches (30 cm.) square or larger. So the Pratt Company ivory products factory in Connecticut—perhaps unaware of (or in competition with) the earlier inventions by Pape and by Cheverton—devised such a machine. Designed and built by two of its machinists, Benjamin Stedman and Fenner Bush, it could cut great lengths of thin ivory from around a section of tusk rotated around the tusk's longitudinal axis. In 1851, the company exhibited its longest ivory sheet at the Great Exhibition in London, where it was suspended from the dome of the newly constructed Crystal Palace in Hyde Park. Julius Pratt reported that it was 14 inches (35 cm.) wide and 52 feet (16 m.) long.[10]

Unfortunately, that startling exhibit has been lost, even though all of the display items were said to have been returned to the exhibitors. (Nor does this writer know of any comparably long ivory sheet available for viewing now.) Further, as with the Pape machine and then the Cheverton machine, the American machine was lost or discarded too.

The Pratt Company later moved to Centrebrook, Connecticut (under the name Pratt-Read Company), where ivory manufacture so dominated the industry of that town that the local populace changed the town's name to Ivoryton. The company became the world's largest manufacturer of ivory products. At one time or another, it made costume ornaments, combs, men's collar buttons (or studs), cigarette holders, toothpicks, letter openers, nameplates, dice, dominoes, piano- and organ-key veneers, and other utilitarian items. After 1954 it no longer made ivory articles, but its president, Peter H. Comstock, maintained an exhibition of the ivory products once manufactured there, as well as some of the old machines.

As a final note on the presumably independent invention of ivory-sheet cutting machines in different countries, it has been reported that M. Alessandri invented one and exhibited it at the Paris Exhibition of 1855. It had cut an ivory sheet 30 by 150 inches (76 × 380 cm.).[11] It seems strangely coincidental that his machine had cut a sheet exactly the same size as Pape's earlier machine. Also strange is the fact that all four machines (by Pape, Cheverton, Pratt Company, and Alesandri) seem to have no known present location.

Here is a caution concerning the use of ivory veneer: thin ivory is somewhat transparent and will therefore show the grain and color of the wood to which it is glued, thus defeating the purpose of using ivory. Where very thin ivory veneer is applied, a craftsman uses underlying grainless wood of very light color, lest the wood's grain or dark color show through. Also, the use of an opaque white glue is advisable.

Special machines were invented for duplicating an artistic sculpture, especially of smaller size than the model. James Watt, the Scottish engineer and inventor, had been making such modeling machines as early as the year 1800. Cheverton once specialized in reproducing in ivory the busts of well-known persons. Using large marble busts as a prototype, he cut small ivory reproductions by means of a sculpturing machine that he and John Hawkins had invented in 1828 (Fig. 7). The operating principle was a complex adaptation of the pantograph device, which transmits the effect of tracing a two-dimensional figure, so that the physical movement produces a duplicate figure. Theoretically, the reproduction can be smaller, equal to, or larger than the original figure, but larger reproductions introduce the problem of a small movement having to produce a large movement. For sculpture, tracing the large model causes corresponding cutting actions upon the small ivory block. In 1924 the machine found its final home at the Science Museum in London.[12]

IVORY CARVING AS A HOBBY

A reader thinking of starting ivory carving as a hobby may wish to consider one carver's experience, as well as certain related issues. Most of the following ideas are based on the work of Carson Ritchie, a twentieth-century English carver.[13]

1. Ivory carving is not physically hard work, especially if one buys small pre-cut blocks of ivory.

2. As a miniature art, it does not require a large working space.

3. For a hobbyist who wants to carve a few pieces a year, the increasing cost of ivory is less of a problem than the increasing difficulty in obtaining it.

4. The required tools and materials are inexpensive for a beginner—coping saw, file (and wirebrush file-cleaner), sandpaper, abrasive powder, and polish powder. Advanced work requires more tools—carving knives, chisels, gravers, and so on. Ritchie describes the tools and, for advanced carvers, offers suggestions for establishing a workshop and a record-keeping system.

5. Ivory carving holds a high place in the art world. In England, for example, ivory sculptures are regularly exhibited at the Royal Academy and at the Royal Society of Miniature Artists.

6. Because there are fewer ivory carvers now, it is easier to achieve recognition as an ivory carver.

7. Prospective carvers, possibly reluctant to use a substance if it requires that the animal be killed, should note: (a) some extinct mammoth tusks are still being sold; (b) the natural death of tusked animals will always supply a substantial amount of ivory; (c) protected elephant herds overpopulate themselves from time to time in some countries and have to be culled, thus making available an added source of tusks; (d) tusked animals, when killed legally, also supply food, oil, and hide for native populations.

Ivory is becoming less available because of species-preservation laws and international agreements. Among the substitutes, the supply of whalebone is limited by the restrictions applied to the killing of certain kinds of whales. Hornbill is rarely available and is very expensive. Vegetable ivory is available from some lapidary dealers and is a good material for small artistic carvings. (Lapidary dealers can also usually tell where ivory can be legally purchased.) Beginners often work with scrap ivory—tips of tusks, thin pieces from the hollow portion of the tusk, broken or waste pieces, and even poorly-done earlier carvings that have little merit or value. The cost is about half the price of good ivory.

Prospective carvers concerned about the hardness of ivory substance can be guided by Ritchie's listings, with the animals presented alphabetically within the sections below. (Other published listings may differ.)

Soft—African elephant tusk (usually from E. Africa), Indian elephant tusk, mammoth tusk, sperm whale tooth, walrus tusk
Intermediate—African elephant tusk (usually from W. Africa), boar tusk
Hard—Hippopotamus tooth
Very hard—Narwhal tusk exterior

After the carving is completed, all accessible surfaces are sandpapered smoothly; otherwise the marks (from sawing, filing, and tool work) will become noticeable after the polishing. Then abrasive powders produce a finer smoothing. Finally, polish powders bring out a natural gloss.

Any reader attempting scrimshaw work should note that Ritchie doubts whether someone could succeed with only the jackknife that most whaler seamen allegedly worked with. The whale-tooth etchings suggest the use of a sharp point characteristic of a steel style. When a sailmaker's heavy needle was not available, a so-called Eskimo style was made from a nail with the head cut off. A hole was drilled into one end of a 3-inch (7.5 cm.) length of wooden dowel; the hole was then filled with glue, and the blunt end of the nail was pushed down into the hole until only about ½ inch (1.3 cm.) of the nail-point end still protruded. After the glue had dried and hardened, the nail point was finely sharpened with a file.

Can a skilled carver earn a living from this craft? It seems unlikely. Besides the present difficulty of obtaining an ample supply of good ivory, high-quality carving is so exacting and so time-consuming that a carver cannot sell his work at a price that would pay for the time expended, and surely not for the value of the skill and creative imagination. This craft is likely to remain a worthy labor of love for all except a small number of esteemed ivory artists.

V

COLORATION OF IVORY

The changing coloration of most aging ivory is unpredictable and should not be confused with the discoloring effect of contact with metal, wood, skin, or other substances. For example, an ivory necklace or bracelet may become discolored due to its frequently coming in contact with skin oils, perspiration, or dirt.

Although some ivories remain creamy white for centuries, others darken or yellow. Very old ivory may resemble boxwood, chestnut, mahogany, or other woods. Ancient ivory may look like slate, basalt, sandstone, alabaster, and even opal or turquoise. Some art conservators think that ivory's original color and tone would be preserved if the piece were protected from air and ultraviolet light rays. Ivory also dries out over time and thereby loses its natural shine. Some polishing agents may restore the gloss.

DATING AND FAKE AGING

The variation in color of ivories from burnt-out buildings of the past suggested a hypothesis to Norbert Baer at the Conservation Center of New York University's Institute of Fine Arts: perhaps the variation resulted from the differential effect of the degree of heat and drying inflicted upon the ivory; then, after such accidental effects became known, heating may have been done intentionally in later times. To test his hypothesis, Baer baked in an oven samples of African elephant ivory at various temperatures for one hour. The samples were then examined for enduring color changes and other physical properties. The resulting colors are shown in the chart on the following page.

Temperature	Color
300° F/149° C	white
400° F/204° C	light yellow
500° F/260° C	brown
600° F/316° C	brown black
1100° F/593° C	black
1200° F/649° C	dark blue-gray
1400° F/760° C	light blue-gray
1500° F/816° C	white
1600° F/871° C	white

Chemical tests indicated that these colors were caused by a sequence of chemical and physical changes in the ivory.[1] Thus, an unusual color in an ivory carving may not always be due to the effect of age.

Fake aging of ivories for the antique market became a common practice in some countries. The simplest method has been to soak the carved piece for weeks or months in either tea, coffee, tobacco juice, wet hay, or various substances. Perhaps the strangest method was the practice whereby small carved ivories (including chess pieces) were given to turkeys to swallow down so that their stomach juices would stain the surface yellow.[2] Patina suspected as fake may wash off when firmly scrubbed, but that would introduce certain risks noted in Chapter 6, "Care and Cleaning."

Old ivories often develop cracks. Because such ivories are respected for their antiquity, fakes have appeared in which the cracks were made quickly by alternately baking and freezing the carved ivory, or by immersing it in boiling water for a while and then immediately baking it in an extremely hot oven until the ivory begins to crack.

Only professional expertise can possibly cope with the chance of fraud in what appears to be an old carving. Even the experts disagree. Suspect examples have remained on display as genuine in some museums in England, Germany, and France.[3] The same uncertainty applies to some ivories in museums and private collections in other countries. Collectors are concerned about the approximate time of the carving process, not the age of the ivory substance. Carving could not have been done long ago if the substance is determined to be from a more recent time; but frequently new carvings are made from old ivory, and this is not necessarily done with any intention to deceive. Determining the age of the ivory substance requires laboratory testing, is expensive, and can estimate age only in terms of broad periods of time.

Assessing the date of a carving is mainly a matter of art history. It is determined according to the subject matter and style of the item and the carving technique used. Unfortunately, early ivories were rarely, if ever, dated or signed. The dating problem becomes more difficult when we see a carving that honestly duplicated an earlier carving. Following are some examples of variation in experts' dating of ivories. Where furnished in the source text, the expert's name is indicated.

1. The Berlin Pyx: second or third century (Hahn), fourth century (Kugler and Kraus), fourth or fifth century (Westwood), sixth century (Molinier).
2. Munich Gallery's tablet "The Ascension": fourth century (Kondakov), fourth or fifth century (Dobbert), fifth or sixth century (Molinier), sixth to ninth century (Waagen), ninth or tenth century (Westwood), eleventh century (Forster).
3. The famous Throne of Maximian (in Ravenna, Italy): experts differ on whether the ivory plaques were made in the same historical period or in different periods, whether the work was Italian or Byzantine, and whether this throne was made in Egypt or Syria.[4]
4. The Louvre's Harbaville triptych: fifth or tenth century.[5]
5. In Milan, the Sforzesco Castle Museum's panel, "St. Mena Between Two Camels": sixth or eleventh century.[6]

Cracks resulting from a few centuries of aging may be seen in some items in Figure 9.

ARTISTIC COLORATION

Arguments have been raised against artistic coloring of ivory sculpture, especially if a large proportion of the surface is colored. One objection is that the natural creamy color is lost when covered with artificial colors. But the artist may wish to mimic the natural coloration of, say, a bird or outdoor scenery. In painted figurines we see how the person represented was dressed. It should be noted that fine artistic coloring is often semi-transparent and thereby allows ivory's surface texture to show through.

Another objection is based on the principle held by some that two different art forms—in this case, sculpture and painting—should not be mixed. Interestingly, we learn that great Greek sculptors such as Phidias and Praxiteles colored some of their marble creations. Similar practices are seen in works fashioned in ancient Egypt and Assyria, and long afterward in Europe during the Middle Ages. The issue is at best a subjective matter of taste, but painting on ivory as an adjunct, without overwhelming the sculpture, does seem acceptable. Perhaps William Maskell merits the final word, in his conclusion that this matter "must be left to the judgment and decision of the time and country for which the sculptures are made."[7]

The kinds of coloring substances used have been numerous, as have been the techniques for applying the colors. In Japan, for example, the artist mixed gardenia and mangrove bark in a vessel of water, and boiled the carved netsuke in this staining mixture for 24 hours; it was then dried thoroughly and polished.[8] Tan-stained ivory figurines from India have become quite familiar in the West.

VI

CARE AND CLEANING

Rapid changes in temperature may cause old ivory to crack. Indeed, a sudden and extreme temperature change may even affect new ivory in the same way.

Older ivory often shows cracks caused by drying out and shrinkage. For controlling humidity, display cases containing ivories often include small glasses of water, especially during dry weather. Some curators recommend a humidity level of 45 to 55 percent; above 70 percent, molds may develop on ivory surfaces. It is sometimes advised that dried-out ivories may be restored by immersion in colorless gelatin or melted wax. One authority recommended boiling a brittle ivory piece in a gelatin solution, an idea perhaps based on Layard's treatment of the Assyrian ivories; the method had apparently been suggested by a Professor Owen.[1] The present writer urges caution concerning such radical methods; consultation with a museum curator is advised.

If an unpainted ivory has become dirty, some collectors use a soft eraser in cleaning; but this may remove the underlying surface gloss, or may leave an unnatural shine on the erased area (as is often seen on soiled coins that have been cleaned with an eraser). Many collectors use a moist cloth. Some use a moistened artist's brush; a few use alcohol instead of water. Unfortunately, these gentle techniques may not remove heavy soiling. If a badly soiled ivory has no cracks, it can be cleaned with soapy water. The present writer has gone further, especially with inexpensive utilitarian ivory products—immersion washing in soapy water and even lightly scrubbing any stubbornly resistant soiling using a soft toothbrush. Then the piece is rinsed in running water and dried immediately. If washing has removed some of ivory's charming gloss, it can usually be restored by rubbing the piece with a soft clean cloth. Some collectors have used lanolin, gelatin, almond oil, and a variety of vegetable oils for this, but research is necessary concerning the appropriateness of various substances for adding gloss (and for preservation of ivory). Caution is also required with very thin ivories (bookmarks, letter openers, and so on), which, if soaked in water, may curl while drying.

Ritchie cleans ivories in soft water, of which 1 percent is ammonia and 1 percent is liquid detergent. He uses a small sponge or an artist's brush. On thin or delicate ivories, he uses equal parts of water and alcohol.[2] One book by Nancy Armstrong recommends that ivory be cleaned with Prelim cream, made in England but also sold in New York City (by Talas, the supplier who also furnishes her recommended Renaissance Wax Polish).[3]

Although collectors generally try to avoid letting water get into cracks in ivory, some say that, on the contrary, they have intentionally doused cracks with water (or have wrapped cracked ivories in wet cloths for a while), because the water then swells the ivory and thereby closes up narrow cracks.

Wide cracks have sometimes been filled with plaster or similar fillers, without necessarily intending to disguise such repair. Ritchie advises, however, that cracks should be repaired by an ivory carver using a sliver of ivory. Where it is not feasible to carve ivory slivers that can be cemented in, a carver could concoct a filler cement made of ivory dust for the purpose.

Special caution is needed when cleaning painted ivories. Water colors may easily be affected; and old dried-out oil-paint coatings may flake off. A common practice is to test-clean a small inconspicuous painted area; if it does not suffer from that cleaning, the same is tried carefully on the other painted areas that are soiled. Or, as with unpainted ivories, one could try a dry method, such as using a soft eraser.

Another caution concerns jointed ivories, where two ivory surfaces have been cemented together. Old cementings, and even new ones poorly cemented, may dissolve or break open if liquids enter the joints. Such care should also be taken when small inlays, gemstones, or precious metals are attached to ivory. It might be advisable to use a blow-dryer for quick drying.

Problems also arise with ivory attached to other materials, such as those that serve as a base. Because small painted ivory sheets are often mounted in wooden frames, some carvers advise that such sheets should only be kept pressed against the backing, not glued to it; ivory must be allowed to expand or contract with various changes in temperature. The extent of expansion or contraction may differ between the ivory and the glued backing, thus causing the ivory to crack or break away. However, ivory sheets on the top and sides of a box have to be glued into position. Fortunately, the risk is now reduced by the use of flexible glue.

When aging has yellowed or darkened the ivory, some collectors have resorted to bleaching, though this is unwise in the case of painted ivories. Most curators feel that old ivory should look like old ivory. Usually the carving is merely exposed to strong sunlight for a week or more. One writer advised that to increase the effect, the ivory should be dipped in turpentine and then exposed to bright sunshine for three or four days.[4] Another writer suggested that a yellowed ivory be immersed for several days in a solution of one part lime chloride and four parts water, and then washed and dried.[5] Others have recommended soaking unpainted, unjointed ivories in hydrogen peroxide solution. In a treatise on the care of antiques, the author advised against bleaching; but if it is attempted, he says one should, while wearing protective rubber gloves, coat the ivory surface with a stiff paste of whiting (with very little water) and 20-volume hydrogen peroxide. Now the piece is propped up in sunlight, outdoors, until the paste dries out. Then the paste is washed off, and the ivory is dried and covered with a very thin coat of almond oil.[6]

This chapter ends with some observations on conservation and restoration of ivory by one curator, Clayton Holiday, at the South African National War Museum in Johannesburg.[7] Aware of the damaging effects of atmospheric changes upon joints, some great craftsmen of the past chose the two ivory blocks in such a way that the two surfaces (to be subsequently joined together) would have the grain running in the same way, so as to avoid the differential contraction suffered by longitudinal surfaces as compared with cross-sectional surfaces.

Holiday closes up cracks in the following way. He washes cracked ivory with "good quality soap" in warm water to remove dirt in the cracks. For a day or two he keeps the ivory in a plastic bag with wet unprocessed cotton; a few drops of alcohol in the water used to wet the cotton prevent the cotton from growing a mold. Finally, almond oil is rubbed onto the entire ivory carving.

He has sometimes been able to reconsolidate the original composition of ivories that suffered physical change, including partially fossilized ivory. Of course, this is no job for amateurs.

VII
TESTING FOR IVORY

Despite the fact that ivory and bone were used as equals for many kinds of useful products, ivory collectors have long been concerned that their collection is ivory and not bone. Of course, if a substantial section of a tusk has been carved, collectors and dealers have no difficulty in identifying it as tusk, and can usually identify the animal source—elephant, walrus, boar, and so on.

The cost of the substance is not particularly relevant here, because ivory is relatively inexpensive when compared with such art substances as gold, silver, and some precious gemstones. Although ivory has greatly risen in price during the latter half of this century, its relatively moderate cost as a substance is reflected by the fact that it is sold by the pound in most places, and may still be sold by the 100 pound at international supply dealers.

Various testing methods will be presented here. Unfortunately, there are severe problems involved, and no single test is reliable, simple, and inexpensive. Reliability, the most important aspect, is seriously limited by the fact that the present tests are only exclusionary: they may exclude a tested carving from the domain of ivory, but cannot absolutely certify that the substance actually is ivory. Moreover, most kinds of testing have at least some destructive aspect by which the entire carving may become slightly damaged or modified, or a small part must be removed or possibly damaged.

SPECIFIC GRAVITY TEST

Ivory, from the several species that provide it, has a specific gravity (density relative to that of water) of between 1.65 and 1.95 in almost all instances. If the carving floats in water, it is not ivory (assuming it has no hollow interiors). We can fairly reliably reject as non-ivory any carving that has a specific gravity outside the 1.65 to 1.95 range. But this method has related problems:

1. Few collectors or dealers would buy the equipment for determining an object's specific gravity.
2. If the carving has attachments of other materials (for example, wood, metal, gemstones), we cannot determine the specific gravity of the alleged ivory part alone.
3. Very few ivory collectors or dealers would allow the carving to be immersed in water, especially if the carving is colored or jointed.
4. Some kinds of bone and several kinds of artificial ivory are in the same specific-gravity range as ivory. So, as mentioned before, this method cannot establish that a carving is ivory.

CHEMICAL TESTS

There are at least two possible methods here. One technique makes use of a unique and noticeable chemical reaction to a drop of some liquid chemical applied to a small surface area; ideally, the reaction is simply a distinctive color change that does not leave a permanent mark. For example, if a drop of diluted sulphuric acid develops a pink spot that can be washed off, the substance may be vegetable ivory from the ivory-palm nut. What is needed is a positive test—a visible, harmless reaction that identifies a carving as ivory.

A second technique requires removal of enough material to allow chemical or physical analysis that could identify ivory. A research laboratory might do this, but it would be expensive, and how many collectors or dealers would allow removal of some material?

BURNING TESTS

A favorite among some antique dealers because of its simplicity, this method requires firmly pressing the point of a red-hot needle into an inconspicuous spot. (The other end of the hot needle is held by gloved fingers or a pair of pliers.) If the red-hot point penetrates the surface, the material is claimed to be neither ivory nor bone. But if the hot point does not penetrate the surface, the material may be ivory or bone, or some other hard and dense material. The burning may leave a tiny black mark, and most carvings do not have an inconspicuous spot for testing. The method can hardly be considered a reliable test for ivory.

A related test is based on the fact that burning will cause a characteristic odor in different substances. With experience some testers have apparently been successful in recognizing ivory or bone in this way. Instead of burning the tiny spot, some testers apply a miniature grinding-bit on an electric drill from a hobby craft-kit; there is said to be a subtle but characteristic odor emitted.

MICROSCOPIC TEST

Although some writers suggest that even a low-power magnifying lens can identify ivory, the fact is that few collectors or dealers are able to do so. The attempt is even uncertain with a microscope at a power magnification of 100 times. This is especially true if we have to examine the carving in solid form, but microscopic identification of ivory is uncertain even when we remove a thin shaving for examination. With experience we can more easily exclude a particular carving from the realm of ivory.

It is likely that the newer kinds of highly sophisticated microanalysis apparatus will succeed in identifying ivory. The testing cost for each carving will be a serious limitation. Particularly for utilitarian products, the cost would exceed the value of the product, even if it is identified as being made of ivory.

LIGHT-REFLECTION TEST

In 1938 Williamson wrote that a "quartz lamp violet-ray apparatus used with a specially prepared screen of Woods' type, but differing somewhat from it in having higher percentage of oxides and nickel, reveals fluorescence in ivory that would appear to be sufficient to determine to what group the ivory belongs." Fossil ivory from the Bering Strait gave a "rich velvety purplish brown fluorescence." A walrus-ivory tusk gave a whiter glowing effect. Elephant ivory from Africa and Asia gave a "clear bright fluorescence . . . with a distinctly blue tinge." But a fossil-ivory tusk from Alaska's Firth River bank did not give the earlier-mentioned velvety purplish brown effect. Therefore Williamson suggested that microscopic analysis be used too.[1]

Ritchie, aware of Williamson's research, believes that this kind of test can only be suggestive, not conclusive; it depends on the particular sample of substance in the carving, as well as the kind of apparatus used for the test. He concluded that the difference between some of his findings and those of Williamson may be due to his use of a standard ultraviolet lamp, whereas Williamson had used a special type of filter that had a higher than usual percentage of oxides and nickel. Ritchie's findings are shown in the following chart.[2]

Substance	Reflected color
African elephant ivory:	
soft	Purple (or blue) and white
hard	White or very white
Indian elephant ivory	Very white with purplish buff streaks
Siamese elephant ivory	Purple and white
Mammoth ivory	Yellowish white
Walrus ivory	Buff and white, with some purple
Narwhal ivory	Buff and white
Hippopotamus ivory	Bluish white and some purple
Wild boar ivory	White, buff, and purplish blue
Warthog ivory	Blue, buff, and pink
Sperm-whale tooth	White, blue, and buff; sometimes yellowish
Bone	Purplish mottling with spots
White plastic	White, "rather like ivory"

Making the matter even more uncertain, one investigator reported that vegetable ivory showed a fluorescence effect similar to (though weaker than) the glow shown by ivory.[3] A different investigator reported that newly cut ivory reflected a purple color, whereas old ivory reflected a yellow tone.[4] Although those latter results may somewhat resemble the general nature of Ritchie's findings—commonality of purple or blue light reflection for ivory—it seems obvious that the light-reflection test might introduce more confusion than contribution in the attempt to identify ivory.

SIMPLE TESTS

In recent years it has been suggested that collectors and dealers, lacking the equipment and expertise to do their own testing, should rely on simple methods, even if of limited reliability. One proposal is based on the relative hardness of ivory. Ritchie indicates that most plastics are harder than ivory and will therefore resist scratch marks from a knife. But here again there is a destructive aspect, as well as uncertain reliability. Many writers, collectors, and antique dealers claim that they can recognize ivory by visual inspection alone, especially with the aid of a magnifying lens. Thomas Penniman wrote a short illustrated treatise about this matter. Basically, his method utilized the descriptions of ivory and its substitutes presented here in Chapters 2 and 3. In the work, he illustrates various specimens from the Pitt Rivers Museum in Oxford, England, where he was the curator.

Some ivories show the familiar criss-cross pattern of intersecting lines on the cross-sectional surface. This is a simple and fairly reliable indication of ivory, but unfortunately too few ivories show this grainy effect.

The last word on this matter may have to be that for the present time at least, most of us must rely on our own past experience with the physical appearance of ivory in whole-tusk form, in small blocks cut from the tusk, and in carvings known with certainty to be made of ivory, as well as on our familiarity with the physical appearance of various substitutes.

PART TWO

THE USES OF IVORY

Any classification system for ivory products will be somewhat arbitrary; this is true for the twenty-one specific and two general categories presented here. An item may be listed under more than one of the twenty-three categories if that item is relevant to more than one grouping, but usually it will only be fully discussed in the most relevant category. Where one grouping is general (such as "Containers") and the other is specific (such as "Personal Grooming Articles"), the item will be discussed under the specific category.

"Containers," the first of the two general groupings, will introduce the historical practice of using ivory for making receptacles. It will treat: (1) particular kinds of containers for which no specific category is provided here; and (2) containers intended for a variety of things, rather than those with a specific use. The second general grouping, "Miscellaneous," will treat those ivory products that cannot be classified under the other categories.

Although a very large number of tools (broadly defined) were made at least partly of ivory, a separate general category for them has not been provided; it would have to include too many examples that are more meaningful in a specific-use category (such as "Cloth and Clothes Making, Embellishment, and Repair"). Any unclassified tools will be treated under the miscellaneous grouping. Similarly, handles of ivory cannot be presented in their own separate general grouping. Since the handle itself can hardly reflect the function of the entire product, discussion of particular articles with ivory handles will be found in the most relevant specific-use category. Some handles, which cannot be classified under such a category, are discussed under "Miscellaneous."

The cross-referenced Index (p. 229) makes it possible for the reader to rapidly locate any particular item of interest, without regard to the item's category classification. Products made of ivory substitutes have not been included, unless the same product has sometimes been made of ivory too. To include unique uses of substitutes (that were never made of ivory) would go beyond the manageable limits of a single work. A major exception will concern bone, accepted by many as the natural equivalent of ivory. Some consideration will be given to products made uniquely of bone when ivory was in short supply and too expensive. Also considered are some products made of large whalebone because ivory pieces would usually be too small. Where no photographs or drawings of ivory products were available for the descriptive sections, a bone substitute may be shown.

Something should be said about the surprising differences in the amount of information presented concerning the various ivory products. In preparing what is probably the first encyclopedic treatment of all known ivory uses, I soon discovered that the venture is similar to an attempt to convert a dictionary into an encyclopedia. Providing the historical and cultural background is not yet possible for most of the more than three hundred different uses of ivory that are listed and described here. Undoubtedly, many of them have never before been discussed in print. Other past uses have likely been referred to in sources that must await discovery through an extensive and detailed search in libraries, newspaper files, invention patents, and personal letters that may have survived. It is hoped that providing the known background of as many ivory uses as possible may spur some readers to communicate other background information personally known to them, or which becomes known some day, by way of laborious research.

Photographs of the distinguished museum pieces of the world were not available for showing here; nor of course was it my aim to present a treatise on museum art in ivory. With its major emphasis on uses of ivory, there is presented here a representative selection of illustrations of ivory products that were made for actual use and which have now come within reach of ordinary collectors with ordinary budgetary means for building an ivory collection. In each instance where a particular ivory product or bone equivalent was unavailable for photographing, it is hoped that the description will clearly indicate the nature of that product. For those readers who might nevertheless wish to see what the article actually looks like, reference (wherever possible) will be made to the ones shown in specialized publications in public or university libraries that may be conveniently accessible to the reader, or which may be obtainable by the interlibrary loan system. If known, the museum of last ownership will also be indicated. An extensive list of world museums with substantial ivory collections is presented in the Appendix (p. 205).

1. Vegetable-ivory products. (a) Ivory-palm nut: two American Indian heads; bust of Bolivar; spin top; two buttons. (b) Betel nut: whole nut; cut piece showing cross-section; key ring with nut toggle with colored glass insets. (c) Black button of tree rubber and synthetics, stamped "Goodyear 1851, I.R.C. Co."

2. Persian miniature of polo players painted on ivory, $3\frac{1}{8}$ by $2\frac{1}{8}$ inches.

[65]

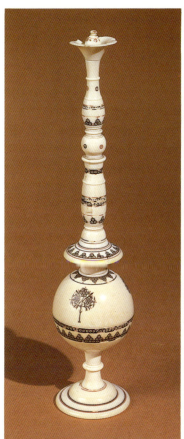

3. Machine-turned decorative stand, $8\frac{1}{2}$ inches high.

4. Orchid-design bracelet-and-earrings set.

5. Netsuke. *(Left to right)* Trained monkey with his master; rear view of skull and snake, showing two holes for the cord (see Fig.11); five monkeys harassing a demon; God of Knowledge with a young reader.

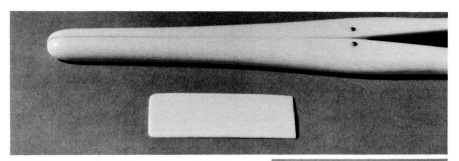

6. Grainy appearance of some ivory. *(Top)* Lengthwise cut surfaces of a glove-finger stretcher and piano-key veneer; *(Right)* Cross-sectional cut surface of a napkin ring.

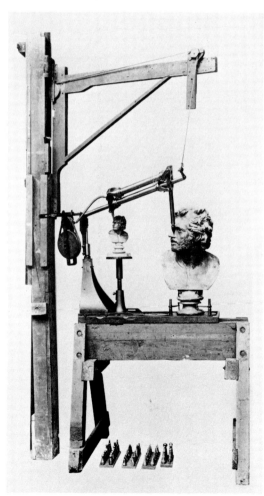

7. Cheverton's reproducing machine.

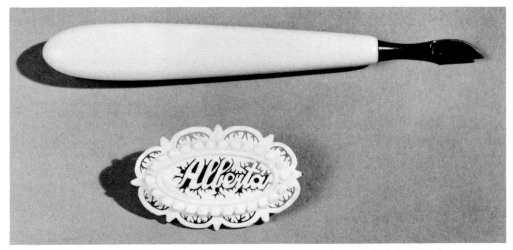

8. Artificial ivory products: manicure instrument with Celluloid handle showing imitation longitudinal grain; plastic brooch sold as French Ivory in New Orleans (1941).

9. *(Top)* Three European carvings of heads; note the vertical cracks due to aging. *(Center, left)* Elder wearing a turban; *(right)* ivory-tower retreat from China. *(Bottom)* Plaque from India.

10. Erotic die with mythical male on female.

11. *(Left and upper right)* Janus head carvings; the front and back sides are the same, as shown in the mirror behind the head on the left. *(Center)* Singapore ball. *(Bottom)* Ivory netsuke, snake on skull.

12. Miniature ivory carvings. *(Top left)* Replica of binoculars. *(Bottom left)* Set of ten animals of ivory and their hollow-seed container from India. *(Center and right)* Intricately turned outer and inner tubes; removed screw top shows a crown insignia.

13. Whale-tooth scrimshaw work, front and back views, 5 inches long (1853). A Great Northern Rorqual (see inscription) is a finback whale.

14. Puzzle ball before embellishment, with eight spheres; the outer sphere is 1¾ inches in diameter.

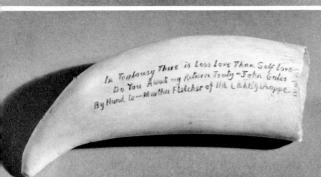

15. Magnifying lens with ivory handle 9 inches long; sandalwood bookstand with ivory inlay.

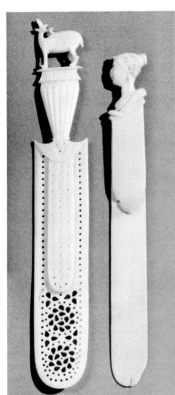

17. Book covers, 4 by 6 inches.

16. Pierced bookmark with animal figure; bookmark with woman's head (the face and extensions have been worn down through use).

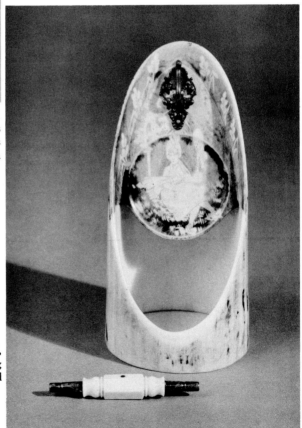

18. Decorated ivory watch-stand, $4\frac{1}{8}$ inches high; watch winding key with ivory handle and metal winding elements at each end.

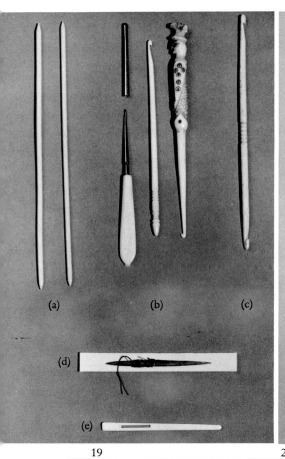

19

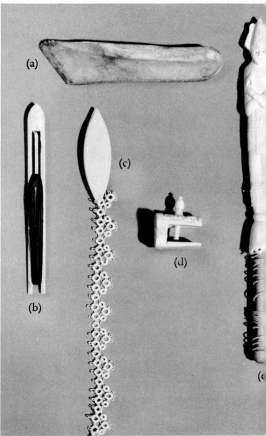

20

21

[72]

(Facing page)
19. (a) Pair of knitting needles. (b) Three crochet hooks. (c) Double-hook model. (d) Eskimo notched sewing needle; thread is attached at center. (e) Ribbon threading needle.
20. (a) Animal-hide scraper made of bone, used by Eskimos. (b) Bone net-repair shuttle. (c) Tatting shuttle of ivory with a section of fabricated lace. (d) Sewing bird. (e) Cotton spinner with ivory handle.
21. Pair of Chinese cases with etched mother-of-pearl discs $1\frac{1}{8}$ inches in diameter; carved three-section container, stained brown.

22. *(Top)* Pair of purse frames with fastener from China. *(Bottom, left to right)* Stained snuff bottle and stopper-spoon from China; snuff bottle with relief carving from Taiwan (back side is painted); round snuffbox with brass emblem on hinged cover.

23. Chinese box; round container with carved lift-off lid.

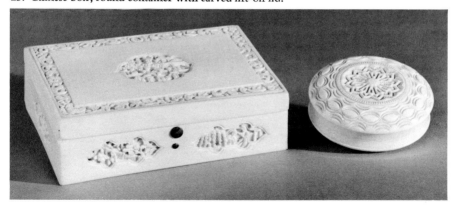

25. Japanese candle holder.

24. Chinese vase, 5¾ inches high.

26. Necklace, bracelet, and earrings set.

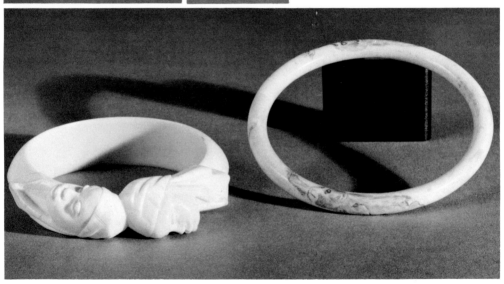

27. Carved African bracelet; monkey-design bracelet from Southeast Asia.

28. Calling-card cases. *(Left to right)* Chinese case; case with stag and filigree work; case with brass horseshoe and stones of turquoise.

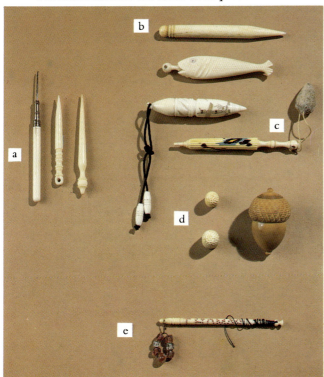

30. Plaque bust of Dwight Eisenhower, 3 by 4 inches.

29. (a) Ivory bodkins. (b) Basting-thread removal pin. (c) Three needle cases. (d) Small thimbles and thimble case. (e) Decorative thread spool.

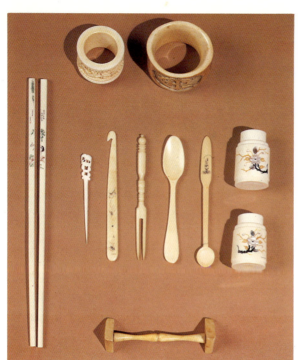

31. *(Top)* Napkin rings. *(Center, left to right)* Pair of Chinese chopsticks; cocktail pick; Japanese orange peeler; pickle fork; two small spoons; salt-and-pepper shaker set. *(Bottom)* Knife rest.

32. *(Top to bottom, left to right)* Toothpick holder; toothpick with pocket case; gold toothpick with ivory case; toothbrush with ivory or bone head; infant's teething ring; doctor's tongue-depressor; bottle vinaigrette and initialed ivory case; round vinaigrette, scalpel with ivory handle; Japanese design pill box.

1 ⊛ ARTISTIC IVORIES

In Chapter 1 (p. 15) we discussed some of the historical aspects and phases in which ivory art arose and flourished. We can see in museums the surviving fragments of prehistoric ivory tusks and bones bearing early man's sketches of certain animals of those times. Before 3100 B.C., Egyptian craftwork is seen with animal figures drawn upon numerous kinds of ivory articles. After 1200 B.C., from the early years of the Assyrian Empire, we have fragments of ivory pictorial panels and figurines. Ancient Phoenician artists began to embellish their carved ivories by adding painted designs and encrusted gemstones. Before 1000 B.C., China was contributing small ivory figurines and artistically decorated useful articles. Ivory plaques, which could rival carved figurines as pure art, often took on a secondary, utilitarian role as, for example, the panels of boxes, doors, and screens. By the beginning of the Christian era, if not earlier, India was producing ivory figurines.

As mentioned earlier, Phidias in ancient Greece produced huge sculptured masterpieces that used plain ivory pieces for the face and hands. His colossal statues of Athena and Zeus have entirely disappeared. Also no longer to be seen, and of greater merit as ivory art, are the ivory statues said to have adorned the ship of Ptolemy II in Egypt. Although classical writings refer to numerous specific examples of ivory statuettes, only a few have survived from early times. Only from the fourteenth century onward do we see a reasonable number of high-quality ivory statuettes remaining, almost all of which are of religious figures.

Apparently it was the Christian devotional ivories that achieved the greatest heights of expressive feeling. They were representational of biblical events and experiences, primarily from the New Testament. The ivory art-forms included panels, plaques, diptychs, statuettes, and crucifixes. As might be expected, the representations of Christ especially—as well as personages closely associated with him or his teachings (Mary and the Christian martyrs, for example)—were often intensely subjective and deeply compassionate. Examples can be seen in many of the art museums of Europe.

Much of the great sculptural art in ivory, both religious and secular, was derivative; often classical paintings and sculptures served to inspire the small ivory art that we now admire. The most frequent examples of such derivative ivory art are from scenes in the life of Christ—for example, the Nativity, Last Supper, Crucifixion, and Descent from the Cross.

In early Christian times, large sculptures in marble or other stone were sometimes placed in a limited architectural space such as a niche, or beneath an arch or canopy. Later, this inspired the practice of carving small ivory panels within an overhead architectural frame. Especially in the instance of the small panels in ivory diptychs and triptychs, the carved figures of people and scenes are sometimes shown beneath arches or canopies.[1]

The curvature so often seen in tall ivory statuettes—inevitable because of the natural curved form of almost all long tusks—became so conventional that ultimately even the stone statuettes that filled the niches of churches and cathedrals of the thirteenth and fourteenth centuries were sometimes intentionally fashioned into a gently sloping form. The same was done for some silver statuettes.

In the seventeenth century, the sculptors' names began to appear on some artistic ivories. Unfortunately, by that time the era of great sculpture in ivory was gradually coming to an end. Identifiable was Andreas Faistenberger (1646–1735), the great carver of ivory crucifixes. Also famous was Simon Troger (1693–1769), whose "Judgment of Solomon" is considered one of the finest works in ivory of the eighteenth century; it is now in the Victoria and Albert Museum in London.[2] Few earlier names are known today.

Chapter IV, "Cutting and Carving," described how miniature painted portraits on small ivory squares and rectangles became popular in Europe by 1700. Then, carved ivory portrait medallions became popular and remained so until almost the end of the eighteenth century. David Le Marchand (1674–1726), who had already achieved fame in France, was the most notable of a number of eighteenth-century foreign carvers who established workshops in London for carving ivory portraits commissioned by the aristocracy and the wealthy. Also, Le Marchand carved ivory figures of Isaac Newton, Christopher Wren, George I, and mythic figures such as his "Venus and Cupid."[3] A large number of portrait medallions can be seen in the British Museum.[4] Around this time, miniature paintings on ivory were also being produced in the Middle East; Figure 2 shows a later example from Persia.

The fully three-dimensional carving (statuette, bust, or small figurine) and the high-relief carving are perhaps the most impressive in revealing the potential of ivory substance as a medium for artistic expression and verity. Figure 9 shows an ivory head (with glass eyes) of a boy, carved in Italy at least two centuries ago. This plate includes two other European ivory carvings of heads, as well as a high-relief carving of an elder wearing a turban, possibly used as a paperweight (origin uncertain), and an Indian plaque, "Young Buddha Comforting a Bird Shot Down." Figure 30 shows an excellent modern example of high-relief carving: a finely carved plaque bust of Dwight Eisenhower as he appeared in the interim between World War II and his subsequent election to the presidency.

In the West, the carved Christian figures were representative, as were a large variety of legendary and mythic figures. In India, for example, probably the most appealing representation was that of the god Shiva, an appeal accentuated by the additional arms. Also, in China a notable portrayal was that of Kuan Yin, a graceful female figure.[5]

Near the end of the last Chinese dynasty (Ching, 1644–1911), rectangular plaques—about 4 by 6 inches (10 × 15 cm.)—were carved with finely detailed designs on one side and microscopic writing (requiring the viewer to use a magnifier) on the reverse side. An entire classic text of 20,000 written characters sometimes appeared on that small surface. Some artists accomplished both tasks themselves: the pictorial representation and the corresponding written account.[6] Also prominent were the fine Chinese structural masterpieces—small pagodas, temples, tea houses, and pleasure boats—usually with surprisingly delicate details. The pagodas and temples, and perhaps even a park scene showing families with their children, were often carved out of a single small block of ivory.[7] Figure 9 shows a small simple example of an "ivory tower" retreat.

Ivory art was occasionally directed toward curiosities. A striking but infrequent form was that of erotic art, primarily a minor speciality of French carving. Figure 10 shows a die of uncertain origin with a mythic male on female in flagrante delicto.

A peculiar subject for carving is the so-called Janus head, with faces at both the front and the back. Presumably, those carvers were unaware of the mythical Roman god, Janus, whose two faces enabled him to look forward and backward at the same time. Figure 11 shows two examples. Interestingly, sometimes one face was male and the other one female.[8] A similar male-and-female face combination is seen in a type of Japanese netsuke in which a movable face can present either a male or female representation. A different kind of curiosity, shown in the same figure, is the Singapore ball, with retractable ivory spikes. Some have thought that it may have been intended as a skin stimulator (see Puzzle Balls and Similar Curios, p. 85).

A grim kind of ivory curiosity was the memento mori (reminder of death) —a small ivory skull (see Figs. 5, 11). A number of German carvers seemed to enjoy furnishing these grim reminders of human mortality. In Nuremberg, Christof Harrich specialized in carving them during the second half of the sixteenth century. To increase the macabre effect, some of the craftsmen carved worms or snakes crawling out of natural openings such as the eye socket. Examples of memento mori can be seen in the Museum of Decorative Arts in Paris[9] and in the British Museum.[10] Occasionally, an entire skeleton was carved in miniature.[11]

Diametrically opposed to such grim reminders, we can see some ivory sculptures that celebrate the vitality and joy of life. The Milwaukee Public Museum has a nineteenth-century Chinese carving of a proud father admiring (or pretending to admire) the acrobatic performance of his two young sons.[12] Many netsuke reveal a delicious Japanese sense of humor.

Some reference has previously been made to ivory work that emphasized craftsmanship rather than artistic spirit. Figure 3 illustrates an example—a turned ivory decorative stand. Figure 12 illustrates craftsmanship in miniature carving: a tiny replica of binoculars $\frac{7}{8}$ inch (22 mm.) long, with see-through magnified views of the Catskill Mountains; a set of ten minutely carved animals, each $\frac{1}{8}$ inch (3 mm.) wide, with their container, a hollow seed, $\frac{3}{8}$ inch (9 mm.) wide; and an intricately turned tube, $3\frac{1}{4}$ inches (82 mm.) long, with a slender $1\frac{3}{4}$-inch (44 mm.) tube inside it (inserted loosely by way of the screw top of the larger tube), and with a barely visible, decoratively turned needle-thin shaft inside the smaller tube. That shaft was probably carved separately and then inserted and glued inside the smaller tube; the shaft is now broken in the middle. References are furnished for illustrations of elaborate and complicated examples.[13]

Art Deco Sculpture Much of the following account of Art Deco sculpture is based on a work by Victor Arwas.[14] The term "chryselephantine" originally referred to statues of gold (Greek: *chrys*) and ivory. In 1897 the Brussels Exhibition provided a separate section for chryselephantine ivories, but the term was broadened to encompass ivory combined with any other attractive substance—silver, bronze, onyx, marble, fine woods, and so on. Critics complained loudly that any carving for which the ivory had to be combined with another substance in order to show a full effect should be considered only as objets d'art, not as sculpture. But just a few years later the International Exhibition of 1900 in Paris presented more of these works; then the same was done in London, Vienna, and Munich.

Admittedly, these multimedia productions were, as Arwas says, "precious in conception and execution, a curiosity to be attempted occasionally rather than the norm." A viewer who comes without a limiting preconception of what all sculpture ought to be can hardly help being impressed by the expanded scope offered to ivory by its combination with other substances. He will also be delighted and charmed by these productions when they are at their very best. For example, we see dancers who strut, pirouette, kick, and sometimes even show off shamelessly. The vigor is often astounding. There are also shy and demure maidens; there are heroic figures, as well as entertainers recognizable as the popular stars from cinema and music hall. Athletes are presented in action, and ordinary sports participants too. The use of jewels and metal makes rich and lavish costuming possible. In no other form of ivory work can the joy and vitality of life be seen so well as in these early twentieth-century figures.

Although fresh gleaming ivory made lifelike the exposed flesh parts (and sometimes the entire figure), the base of the Art Deco production was almost always an important part of the composition. The preferred material for supporting the figure was either onyx or marble. The base was usually carved elaborately, either round or with several sides, perhaps in varying horizontal layers, or as a stepped pyramid or some other solid geometrical shape.

Whereas the sculptures of this kind in the 1890s had been expensive objets d'art for any aesthetes able to afford them, the works of the 1920s and 1930s were less expensive decorative objects for the middle class and served as ideal presentation gifts for weddings or retirement. Ultimately, their successful appeal brought about a multitude of cheap imitations, with the ivory now replaced by Ivorine or other ivory-colored substitutes. These inevitably cast a shadow upon the authentic productions and led to the virtual extinction of the art form. For years after that, the major London auction houses were no longer willing to include them in their specialized sales.

But in time these unpretentious and highly decorative multimedia works have re-established their charm and appeal and become a favorite for a new generation of art collectors. No mere description can do them justice. To be appreciated, they must be seen, even if only in a photograph. The best source is the fully illustrated treatise by Arwas mentioned early in this section.

Netsuke Because the kimono has no pockets, it became customary in Japan to suspend small personal articles—purse, seal case *(inro)*, writing materials—from a cloth belt, or *obi*, worn around the waist. A short string was attached to the suspended article; the other end of the string was attached to a toggle or netsuke (pronounced *ne-tske*).

Originally, the toggle was simply a piece of plant root *(netsuke,* root attachment), or a small piece of wood, bone, or other hard substance. Later, when aesthetic appearance began to play a cherished role, the holes for the string attachment were cut in a looping tunnel-like fashion, with both openings on one side of the netsuke. The hole openings were cut on the side to be worn next to the body, out of sight. A knot prevented the string from slipping through the netsuke. When a carver wanted to make the knot as unobtrusive as possible, and to avoid it rubbing against the body, one hole opening was made large enough to contain and conceal it. Sometimes a netsuke had no holes cut in it if it had been carved in such a way as to have a small aperture through which the string could be passed, such as a curved arm touching the figure's body. Some containers were attached by a double string, so that a large bead *(ojime)* could slide along the parallel strings to tighten or loosen the cover of the container, or keep closed a multisection container such as a seal case.

Although the toggles and beads were originally made from common substances, the later ones were made of ivory or gemstones. Unembellished at first, the toggle and bead gradually evolved into the netsuke and the ojime as we now know them. They have become miniature artistic sculptures that exhibit not only high technical craftsmanship but also, in the netsuke, imaginatively exquisite and creative conception. By the end of the 1600s, ivory was being used frequently from elephant, fossil mammoth, narwhal, walrus, hippopotamus, wild boar, and sperm whale (teeth). For containers, even the ojime bead became a thing of beauty. Usually made from agate, amber, coral, jade, or other gemstones, it was also made from ivory. When crafted by recognized artists, the ojime and the netsuke were often signed.

The subjects for the netsuke might be animals, fruits, vegetables, one or more persons at work or play, mythological figures, or numerous other representations. Sometimes the theme was humorous, depicting domestic or legendary incidents; or it might be grim, for example, a skull netsuke, or even a skull with a rat inside, peering out of one of the hollow eyesockets. The theme of a situational netsuke might illustrate a moral precept, such as heroism or loyalty. Other unusual kinds of netsuke will be mentioned later. Examples of netsuke are shown in Figures 5 and 11.

Because the toggle had to be handled continually, and also feel comfortable when it pressed against the body, it could have no sharp edges and no protruding points that might break off. Nor should it catch onto the cloth threadwork of the kimono. And because the mobile netsuke might happen to be seen from any position, the thoughtful artist carved and decorated the entire surface area. Even though most of the later netsuke were carved in such a way that they could stand up on their bottom surface if used for display, rather than as a toggle, that hidden surface was carved and decorated equally well. Concerning the time and patience devoted to the planning, carving, and decorating of the finest netsuke, Yuzuru Okada tells us that "It was not rare that a month or even two months were spent in making a single netsuke."[15]

Netsuke should not be confused with *okimono,* a similar but often larger sculpture, intended only for display. Because they were usually larger and might also be fragile or have protruding points, they would not be suitable as a clothing accessory. Therefore they had no holes for a cord and were frequently unfinished on the bottom (standing) surface.

Decorative toggles seem to have been introduced around the 1300s. Truly artistic netsuke appeared in the late 1500s or early 1600s (surviving early examples are rare). Until somewhat later, these works of art were not signed by the artists.

Unfortunately, many of the later netsuke carried forged signatures of famous carvers (the forging of eminent carvers' names has been reported to have occurred as early as 1781).[16] As with many other art forms, the detection of forgeries has become a seemingly hopeless task. It should, however, be noted that many later carvers had innocently and honestly intended that the signature should indicate the artist whose type of work had been copied. On a lighter note concerning signatures, one scholarly collector had been certain that a particular netsuke in his collection would be signed, but he was unable to locate any signature. More than twenty years later, he happened to discover it, minutely scratched inside the cord hole.[17]

Many carvers devoted their lives exclusively to the netsuke art-form, and some confined themselves to a single theme. One artist, for example, always carved netsuke that showed a monkey with his human trainer; and yet each such carving was different in some way.

The painting of ivory netsuke began around 1750, whereas staining— usually in brown dyes from berries—began after that century. Inlaying with various gemstones also started around 1750; by the next century it became so extensive that critical opinion began to scorn it as uninhibited vulgar display. Carvers' study and research on techniques led to many refinements. For example, in order to simulate cloth in the garments worn by a figure, that part of the ivory surface was lightly roughened with a special tool. Also, although there were already numerous techniques and substances for staining an ivory, one more was added when an ingenious carver began to use incense smoke for imparting only a soft tint to an ivory carving.[18]

During the eighteenth century, the increased popularity of pipe smoking led to a great increase in the demand for netsuke used for supporting the pipe case and the tobacco pouch. The great golden century for netsuke ran from approximately 1750 to 1850; but some critics constrict the period to a so-called Golden Age, from 1780 to 1850. The practical need for netsuke toggles lessened as cigarettes began to replace the pipe, thus eliminating the need for the pipe case and tobacco pouch and their supporting toggles. Then Western style clothing replaced the kimono and provided pockets. With this, the utilitarian role of netsuke sharply declined.

Until the United States helped to open Japan for commerce in 1854 during Commodore Perry's visit, few netsuke had ever left Japan. When the shogun (military ruler) was deposed in 1868 and the emperor was installed as supreme ruler, the dispossessed nobility, thus deprived of their revenues, began to sell their art treasures. In the next twenty years, thousands of the finest netsuke ended up in European and American shops. In England, one collector by the name of Walter Behrens amassed 6,000 netsuke.[19]

Carving of these miniature showpieces continued as an art industry, or at least as a source of attractive and marketable curios; and carvers even continued to include the cord hole that would not be used. When tusk ivory became scarce, old ivory Buddhas and other religious figures (even from Japanese temples) sometimes ended up in sales to netsuke carvers. Also, ivory sawdust cement was pressed into netsuke form as the ivory cement hardened, and this work was done so skillfully that only an expert connoisseur can detect the "falsification."[20] Small Chinese ivory figurines were sometimes converted into alleged netsuke by simply drilling a pair of connected holes in them; the same was done with some okimono, the stand-up carvings.[21]

Ultimately, a few of the great private collections were bequeathed intact to museums, but most of them went to auction sales in the early 1900s. Smaller collections went to antique shops and specialized art shops. Thus were the collections broken up, to the eventual benefit of a second generation of netsuke collectors. In recent years it has been reported that old exceptional netsuke have been auctioned off at prices of $10,000 or more. At possibly a record price for a single netsuke, one was sold for $27,000 at a London auction.[22]

Unusual kinds of netsuke may be of interest to the reader. Some had movable parts—tongue, jaw, or eyes; the eyes might even pop out about ¼ inch (6 mm.). An even more surprising netsuke of this type shows a rotten pear with a small hole on the surface; by moving the pear back and forth, we see a loose worm coming half-way out and then disappearing inside again.[23] On another kind, the head could be turned around; or the head might have two different faces, either one of which could be turned to the front. It has been stated that the carver Shogetsu introduced the reversible face. These were based on Kabuki stage-portrayals where an actor uses a mask to change into another character.[24] The American Museum of Natural History in New York City has an ivory okimono that shows a seated figure wearing a lion head-covering; five faces can be presented, rotating vertically. Also, there have been trick netsuke, such as a figure with a rounded and weighted base, so that the netsuke will always return to an upright position if it has been laid on its side or knocked over. Some visibly top-heavy netsuke are carved in such a way that the figure can stand on one foot, balanced perfectly.

Rarely, a netsuke had a secondary practical use, in addition to serving as a toggle. Some examples of miniature articles that were either a part or all of an ivory netsuke are abacuses, ashtrays, containers, mariner's compasses, signets (seals), and sundials. These could not often have been made, it seems, because ivory models are very seldom seen; more frequent were the ones made of metal or other substances.

Although the period after 1850 yielded a steady supply of mediocre ivory netsuke for the tourist trade, a more serious interest in collecting fine netsuke has not been extinguished; it even began to increase around 1925. Skilled carvers have again furnished miniatures of very high quality, with new expression of originality and charm. Contemporary netsuke—no longer used as toggles—are often carved without the restrictions that would be imposed by the toggle function. They sometimes have sharp edges or protruding points, and the cord hole may be absent. Such art works have become miniature okimono, that is, simply display pieces. Purist collectors, however, avoid such carvings and do not consider them to be netsuke. Also, there are less of the grotesqueries and other exaggerations of the past. The modern, more cheerful looking netsuke have greater appeal for many of the newer and younger collectors.

The majority of recent netsuke have been made of ivory, rather than of wood or other substances. But the greatly increased cost of ivory now, and the high cost of living in Japan today, allow only a small number of established craftsmen to take considerable time for carving an exceptional netsuke. A modern carver may take from one to four weeks to plan and complete a netsuke of very high quality. A few carvers still take six or more weeks, as in earlier times.

Readers interested in further study of netsuke should refer to the already cited works by Miriam Kinsey and by Raymond Bushell, and to a work by Mary Louise O'Brien.[25] All three books are profusely illustrated.

Finally, here are some of the museums with fine collections of netsuke: Tokyo National Museum; British Museum and the Victoria and Albert Museum in London; American Museum of Natural History in New York City; Museum of Fine Arts in Boston; Newark Museum in Newark, New Jersey; Asian Art Museum in San Francisco.

Puzzle Balls and Similar Curios It has often been said that the puzzle ball, a set of thin ivory spheres carved one within another, first appeared in China during the Ching dynasty (i.e., no earlier than A.D. 1644). However, Berthold Laufer has reported that such unusual productions were made in Canton as early as the fourteenth century and were called "devil's work" balls.[26] Of course, there is no puzzle requiring solution, but it may be puzzling as to how such a thing can be made. Before the artistic embellishment begins, the cutting of the spheres expresses not ivory art but rather ingenuity and craftsmanship. It is not surprising that these carvings were also called magic balls.[27] Although the earliest puzzle balls were apparently carved solely by hand, later ones were produced using machine lathes and special cutting tools which automatically cut away the inner balls, or "shells."[28]

After about six to twelve tapered holes have been drilled through from the outside of a solid ivory ball down almost to the center, the innermost (smallest) ball is cut away first. Then the successive balls, each one a little larger than the preceding one, are cut away in spherical layers. The inner balls are incised, sometimes cut straight through, and often each one has a different design. The outside ball is thicker and it receives elaborate sculpting; added painted effects may show, for example, a Chinese dragon. The present writer has seen one reference which states that the outer ball was cut away first, and then carving proceeded down to the tiny solid ball in the center.[29] For an illustration of the concentric spheres before embellishment, see Figure 14.[30]

What is the record number of carved balls within a single work? It may be the example displayed in 1915 among the Chinese exhibits at the Panama-Pacific International Exposition in San Francisco—twenty-eight balls, made by Li Hsao-yu.[31] It is very difficult for us to even count the actual number of inner spheres when there are more than nine or ten. Imagine the skill of the craftsman who has to reach in with miniature cutting tools and decoratively incise each ball. When one inner ball breaks during the incising, as happens sometimes, the entire production is greatly reduced in value. The broken ball can be removed entirely, bit by bit; but now the smaller ball just below the missing one will rattle loosely within the larger space around it.

By the mid-1600s, Lorenz Zick of Germany was carving these curios. He may have seen an early example that came from distant China. Also, there was a similar kind of craft in Europe. For example, the National Museum in Florence has a work from the early 1600s by the Cypriot Filippo Planzone: a small ivory horse carved inside and visible within a previously carved network container. As in the instance of the spheres, the inside carving cannot be removed.[32] It was mentioned in Part One that Lorenz Zick is reported to have carved another object, similar to puzzle balls: a set of so-called counterfeit boxes, carved one within another. Also, his son Stephan was apparently the first to carve a set of ivory trinity rings (see p. 22). At about the time of Stephan Zick's death, a similar set of three interlocking ivory rings (in a small ivory case) was carved by Gustav Magnus Stenbock (1664–1717), who was born in Sweden but worked in Denmark. This carving, made around 1714, is at the Rosenborg Palace in Copenhagen.[33] The carving of two or more inseparable pieces (usually of wood) has in fact long been a popular craft form, especially in the Orient.

Later, France too furnished puzzle-ball carvings. Although the French

Revolution led to a decline in ivory carving in Dieppe, the early 1800s brought many English tourists there; they began to buy up the accumulated stocks of French ivories. Meanwhile, the imported quaint Chinese ivories—including puzzle balls—were selling so well that the Dieppe carvers soon developed the fine skills of carving the puzzle balls themselves.[34]

A different kind of ivory curio, the Singapore ball, was once mistakenly thought to have been carved externally from a single piece of ivory. It was a small solid ivory ball with anywhere from twenty-five to sixty holes in the outer surface. Inside each hole was a small, sharp ivory cone, loosely trapped. No matter how the ball was held, the underneath surface would show the pointed end of some cones dropping down by gravity. Actually, the large end of the cones had been forced into the holes; so such curios represent simple assembly work, rather than skillful carving from a single piece of ivory. The Singapore ball is said to be based on the appearance of the lotus flower's perforated seed-pod.[35] This kind of curio was also carved by German craftsmen and was called *Stachelkugel,* or "spike ball (Fig. 11).

From Germany came a highly imaginative and skillful production (also of an assembly nature) called a *Contrefaitkugel,* or "counterfeit ball." It has a hollowed-out interior and a small hole in the front and in the back. The craftsman inserted and skillfully assembled minute pictorial items which could be observed through the hole in the front. The ivory stand supporting the ball was carved elaborately and intricately, as were also the pieces attached to the top of the ball. Sometimes a miniature chandelier was hung inside. In some models, draw strings on the outside could raise into view any of several miniature scenes inside.[36] One model had portraits of Empress Maria Theresa of Austria and her consort husband; inside the base there were forty-five miniature ivory household utensils.[37]

Also interesting, but not providing for manipulation, was a Chinese production which was almost entirely a form of interior carving. An elaborately carved palace was carved from the inside substance of the ivory ball. It showed extremely minute chairs, folding screens, and human figures, as well as lanterns and tinkling metal bells hanging from the eaves.[38]

Scrimshaw The term "scrimshaw" is of uncertain origin, with various possibilities having been suggested. Narrowly, it refers to the decorative process and product whereby the lower teeth of sperm whale (a form of ivory) were etched on the exterior surface by seamen, after which the etching was darkened with black ink and sometimes with colors from tobacco juice, berries, or other plant juices. Other marine substances have been used instead of whale tooth, such as whalebone, walrus tusk, and sea shells. Whale-tooth scrimshaw work is shown in Figure 13.

Although it is generally believed that the term was probably derived from an earlier one that referred to the etching process, a more interesting suggestion has posed the possibility that a Captain Scrimshaw may have headed the whaling ship that started the practice. Or perhaps a seaman of that name was the first to decorate an extracted tooth from a whale. The earliest written reference to one of the several forms of the term appears to be the one in the written log of the brig *By Chance* on May 20, 1826. The first use of the term in a printed publication may be the mention of scrimshaw teeth and jawbone of the sperm whale in 1841 in Olmsted's *Incidents of a Whaling Voyage*.[39]

The earliest dated scrimshaw pieces (1665 and 1712) were both found in England, but these were walrus-ivory tusks, not sperm-whale teeth. British northern explorers may have learned about scrimshaw work from the Eskimos, and likely brought home some decorated walrus tusks. American seamen may have first learned about scrimshaw work by way of contact with British whalers.[40]

As regards scrimshaw work on a whale tooth, the earliest known dated example is a tooth with the etched caption "1776 Nath'l Healey," now in the Los Angeles County Museum of Natural History. A whale tooth dated 1798 shows an eighteenth-century naval vessel. For both of these it is of course only a supposition that the dates are the actual dates of carving.[41] But not long afterward, by 1828, we can see authenticated scrimshaw teeth crafted by Frederick Myrick aboard the American whaler *Susan* (built in 1826, lost in the Arctic in 1853).[42] An example can be seen in the Nantucket Historical Association in Massachusetts.

In the days of whaling, scrimshaw became almost uniquely an American art form, prominent especially among New England seamen. In the whaling trade, whale oil was paramount; the sperm whale's teeth were of little practical value. Thus, the teeth afforded the crew a readily available substance to work upon as a relief from the boredom of the many periods of inactivity. Although these seamen often saved their completed scrimshaw pieces as intended gifts for a wife or girlfriend back home, some pieces were bartered for tobacco from a shipmate who did not do scrimshaw work; and it is believed that most scrimshaw carvings were sold at foreign ports of call.

Because only the tip of the whale tooth was smooth, the sides, being ribbed longitudinally, had to be prepared before etching if the scrimshander wanted a smooth surface. He did this by scraping or filing down the ribbing, or by patient rubbing down with a piece of abrasive sharkskin. Then the surface might be sandpapered too. For a final smoothing operation, a slightly damp cloth was dipped in scouring powder or in powdered pumice. Then a clean, dry cloth was used for putting a polish on the entire surface. (In modern times, power tools have accomplished all these operations better and faster for the new generation of scrimshaw artists.)

For the etching process, early whalers seem to have used only a jackknife or a sailmaker's needle. Many of the later scrimshanders made their own so-called Eskimo stile, a sharp nail that protruded from a short wooden peg (the method was described in Part One, p. 48). The advantage of using a sailmaker's needle was the fine and delicate details that could be incised. For example, sometimes a surface was incised under a magnifying glass so finely that a magnifier was also needed in order to see the details.[43] In time, some of the more sophisticated seamen began to bring aboard a small set of dental instruments.

For pictures and descriptive captions, some seamen used their own fancies or scenes they had witnessed at ports of call; others simply traced pictures from the magazines and books that were always brought aboard on these long voyages. Scenes of ships at sea were popular subjects. Probably the most poignant depictions for these seamen, who would be absent from home and loved ones for two or more years, showed their tearful partings or their anticipated joyful reunions. Rarely, instead of etched scrimshaw on a whale tooth, high-relief carving was done, mainly of a human figure; the Nantucket Historical Association has some examples. Even rarer was a solid artistic carving from a whale tooth, an example of which can be seen in the New Bedford Whaling Museum.[44]

With the severely reduced availability of sperm-whale teeth after many countries officially declared these whales to be an endangered species, natural-looking plastic imitations have been offered to buyers without any intention to deceive them. Unfortunately, later reselling of imitation scrimshaw (by some unscrupulous individuals or dealers) labeled the items as "whale-tooth scrimshaw." A final word is sadder yet, though not detracting from the interesting nature of this art form: because whaler seamen were not truly artists, and often merely traced an illustration upon the whale tooth, whalebone, or walrus-tusk, scrimshaw work did not gain high respect in the world of graphic or sculptural art. (The largest U.S. museum collections of scrimshaw work are detailed in the Appendix, p. 205.)

2 ❈ BOOKS AND BOOK ACCESSORIES

By their very nature, books allow only a few possibilities for ivory usages. But broadly considered, some interesting applications have materialized, including in a limited way the subject product itself—ivory books.

[90] Part Two : Uses of Ivory

Books of Ivory   Because ivory sheets can be written on or painted, it was inevitable that at least a few ivory book productions would appear. The Victoria and Albert Museum has a six-page fourteeth-century picture book from eastern France or the Rhineland. The ivory covers are carved and painted with New Testament scenes. The six leaves were originally waxed ivory writing-tablets, but were later painted with scenes from the life of Christ.[1]

Another example is the pictorial tangram book. Tangrams consist of seven flat geometrical shapes, all of which have to be used to form any of several provided figures (see Tangrams, p. 166). A tangram book shows a variety of figures and may also show the solutions. These books, when made of ivory, may have as many as twenty-five small printed ivory sheets.

Book Covers   In some instances, uncertainty exists as to whether the elaborately carved covers of an old book were originally a pair of plaques, or perhaps the two sections of a diptych. Conversely, some original book covers were later converted into separate plaques or a hinged diptych. Usually, only the front cover of a rare or precious book was carved artistically. After about A.D. 800, ivory was increasingly used for making fine book covers for treatises of special merit.

The Vatican Museum has a book cover from the sixth or seventh century, with carved themes from the New Testament.[2] The Victoria and Albert Museum has a very finely carved book cover from the ninth century, showing the Nativity, Virgin and Child, angels, and other figures.[3] Of even greater interest, the British Museum has a fourteenth-century French plaque used as a book cover. In a space of only 6 by $4\frac{1}{4}$ inches (15×11 cm.), there are 30 small panels, 6 in each of 5 rows. Each panel shows a scene from the life of the Virgin and contains groups of up to half a dozen or more persons.[4] By the fifteenth century, book covers of ivory began to appear in India. A pair of simple covers bearing a silver shield within a silver knotted ribbon, probably nineteenth century, is shown in Figure 17.

Bookmarks   Although any plain, thin ivory strip could probably serve as well as the customary strip of ribbon or leather to mark a page, more decorative ivory bookmarks were desired. They were carved so that a short flap extended down from the top; the long page marker would slip down over the top edge of the page and remain in that position (Fig. 16).

Bookstands   Bookstands decorated with ivory were popular in nineteenth-century India and were still being made there at the beginning of this century. Figure 15 shows a sandalwood bookstand, probably made in the early nineteeth century. Its two hinged sides are covered with tortoise shell, overlaid with ivory designs; the base has ivory inlays. With age, many details have dried out and fallen away.

Magnifying Lenses Magnifying lenses were sometimes set inside an ivory ring, the ivory serving mainly as a comfortable holder, but also protecting the fingers against possible sharp edges on the rim of the glass lens. Where a handle extension was desired, ivory was used occasionally instead of silver or other fine materials (Fig. 15).

Reading Pointers Silver and ivory wands have been used for this purpose. The ivory models were often decorated with carved foliage or other designs, and with the pointing end carved in the form of a miniature hand with the forefinger extended. London's Jewish Museum displays several ivory models, varying between 8 and 11 inches (20–28 cm.) in length.[5] Now one would be a rare find for a collector.

Scroll Rollers One of the earliest forms of book consisted of a long strip of either papyrus or vellum, rolled onto a pair of wooden cylinders. Later, the more precious scrolls sometimes had rollers made of ivory. During China's Sung dynasty (A.D. 960–1279), some had ivory knobs attached to the ends of wooden rollers. Very few of such articles have survived.

3 ❂ CLOCKS, CLOCK CASES, AND ACCESSORIES

The measurement of time has made possible a small number of interesting applications for the use of ivory, starting with a clock made entirely of ivory—the sundial.

Clocks: Sundials Long before the advent of the water clock and the mechanical clock, sundials fulfilled the need for a simple approximation of daily time (at least during the periods of sunlight). Some of the finer ones were made of ivory. Leonhart Miller made a portable travel sundial of ivory in 1652; it is now at the Old Town Museum in Aarhus, Denmark.[1] These small pocket dials were a popular item in Western Europe during the seventeenth and eighteenth centuries. French models made of ivory can be seen in Dieppe at the Old Chateau Museum. Collectors now could hardly expect to obtain one.

Clock Cases Large utilitarian products of ivory had a special fascination for many ivory lovers in the past. The problem with a clock case is that a circular section from even the large, hollow base of an elephant tusk can accommodate a clock no wider than about 5 inches (13 cm.), with the exception of the very large, more expensive tusks. Clock cases of ivory were made in limited numbers and are rarely seen now; when the clock no longer worked well, they were probably thrown away. The Anchorage Museum has an ivory clock-case $4\frac{1}{2}$ inches (11.5 cm.) high carved by Alaskan Eskimos.[2] One is also at the Mystic Seaport Museum in Mystic, Connecticut.

After imitation ivory was invented, clock cases began to be made of such compounds as Ivorine, Ivoroid, and French Ivory.

Metronomes A British writer tells of a homemade, and possibly unique, metronome timer. Lescarbault, an amateur French astronomer, defending the accuracy of his timing of the transit of a supposed new planet across the sun's face, used an ivory ball suspended on a silk thread; the swings had indicated time in seconds.[3]

Watchbands For the tourist and export trade, Alaskan Eskimos have made ivory watchbands for wristwatch wearers who wanted something a little different. They also made ivory fobs, that is, handle attachments for watches carried in a vest or waist pocket.

Watch Stands In the nineteenth century, and probably even earlier, India was furnishing ivory-inlaid watch stands of wood for holding upright a timepiece. Figure 18 shows a decorated watch stand (of indeterminate origin) in which a small section of tusk serves directly as the stand.

Watch Winding Keys It came as one of many surprises to the present writer to learn that even an article as unpretentious as a watch winding key might have an ivory handle. Figure 18 shows a winder with an ivory handle. Few were made, and much fewer have survived.

4 ✸ CLOTH AND CLOTHES MAKING, EMBELLISHMENT, AND REPAIR

The fabrication, embellishment, and repair of cloth and clothes utilized a variety of implements made of ivory. In a few instances, as will be seen, bone was also employed, especially where attractiveness may have been considered less important. Before 3100 B.C., Egyptian ivory and bone artifacts for hide preparation included scrapers, awls, and sewing needles.

Animal-Hide Scrapers Ivory or bone scrapers were used mainly for scraping the hide's inner surface free of the attached extraneous fat and flesh. The scraper had to be made keen enough for this task, but not so sharp as to cut into the hide. They were also used for removing the scales from fish, as well as other domestic operations. Many examples of scrapers fashioned by the North American Indians and Alaskan Eskimos have survived (Fig. 20a).

Basting-Thread Removal Pins Tailors sometimes used a large ivory or bone pin for removing the temporary basting-thread from custom-made garments during an early phase of fabrication (Fig. 29b).

Bodkins The earliest known ivory bodkins were found in Egyptian pre-dynastic tombs (i.e., dating before 3100 B.C.). These implements are used for punching holes in cloth or leather before threading or adding attachments. They were sometimes called sewing awls. Some bodkins had a large eyehole at one end for drawing a narrow cord or length of elastic through a hem or other channel on a garment. Bodkins are shown in Figure 29(a); one has a steel point and ivory handle. Bodkins have also been used as removal pins for basting-thread, described above. (See also Threading Needles, p. 95.)

Cotton Spinners At London's Great Exhibition of 1851, an ivory spinning wheel was displayed.[1] Of course, this kind of product required a fairly large number of assembled ivory parts.[2] Figure 20(e) shows a simple but attractively carved cotton spinner which was hand held.

Crochet Hooks and Knitting Needles Very large numbers of knitting needles and crochet hooks have been made of ivory. Usually they were plain, without any attempt at decoration. Although most of the crochet hooks were functional at only one end, some had a hook at each end. Figure 19(a) shows a pair of plain ivory knitting needles; 19(b) shows three regular crochet hooks, one with a steel hook that can be covered with a metal sleeve, another with a decoratively carved handle; and 19(c) shows a double-hook model.

Darning Eggs These were put into the toe or heel of socks to facilitate mending. Occasionally, they were made of ivory. A rare model, made by some whaler seaman, is hollowed; the two halves, when unscrewed, reveal an ivory thimble, a small cylindrical ivory case with pins and needles, and a tiny ivory spool of thread. Another rare model is a double-ended darning egg for mending gloves; one egg-shaped end is small and the other is of medium size.[3]

[94] Part Two : Uses of Ivory

Glove-Finger Menders Serving a function similar to that of a darning egg, the ivory fingers of this aid were of different sizes to accommodate the different sizes of the glove fingers that might need mending.

Measuring Sticks For use in dressmaking by their wives or mothers, whaler seamen occasionally made a measuring stick from whalebone. It resembled a yardstick though not always 1 yard long.

Needle Cases Among the numerous specialized kinds of containers, needle cases solved the problem of housing the tiny slender implements that are easily lost. The better quality cases were often made of ivory in highly varied styles. Figure 29(c) includes three needle cases. The fish model shows magnified views of the 1898 Omaha (Nebraska) Exposition; the umbrella model has an attached pin cushion.

Net Shuttles For repair of fishing nets, a special type of shuttle was employed. A standard model, made of bone, is shown in Figure 20(b).

Seam Rubbers Consisting of a heavy handle and a large head, a seam rubber was used for flattening the raised seam that had been newly stitched on a pieced canvas sail, so that any inadvertent contact with the seam would not wear out the stitching. Whaler seamen made them of ivory or whalebone.[4] Some can be seen at the Mystic Seaport Museum in Mystic, Connecticut.

Sewing Birds The earliest fabric clamps were unadorned, but later ones added charming little figures such as birds, dolphins, dogs, or butterflies. The name "sewing bird" apparently came from the model in which the fabric was held between the upper and lower bills of a bird's beak. The materials used in making the device included ivory (Fig. 20d). The first U.S. patent on a sewing bird was granted in 1853.[5]

Sewing Needles Figure 19(d) shows a simple type of ivory needle, double-pointed, with the thread attached at a notch in the middle. Other ivory needles had an eye-hole (see also Yarn Needles, p. 95).

Swifts A swift was an apparently complex yarn reel—consisting of more than one hundred separate pieces—used for firmly holding a long roll of yarn. It allowed unwinding of the yarn while the yarn was being hand-wound into a compact ball that would later be knitted into an article of clothing. First, a loose hank of yarn was placed upon the vertical swift's perimeter framework, which looked like the ribs of a small umbrella. Then the framework was spread out so as to hold the horizontal roll of yarn firmly (much as a helpful husband might do with his hands spread apart). Swifts were made of ivory, whalebone, baleen, and wood, either solely or in combination.[6] Some can be seen at the Mystic Seaport Museum.

Tatting Shuttles This small elliptical shuttle (with pointed ends) was used for fabricating tatting—handmade lace, fashioned by looping and knotting a single cotton thread attached to a shuttle (Fig. 20c).

Thimbles and Thimble Cases The earliest thimbles in America—of ivory, silver, or steel—came from England. (The American patriot-craftsman Paul Revere made a gold thimble for his wife.) By 1800, American companies were making thimbles of ivory, and then of silver (Fig. 29d).[7]

Thread Spools and Barrels Before the French Revolution, craftsmen in Dieppe made ivory navettes—that is, spindles for silk thread. Figure 29(e) shows a decorative spool (inscribed "Thomas"), which is more attractive than practical, since it lacks provision for an ample supply of thread; it was bought in Hong Kong in the late 1800s. Chinese thread barrels were brought to the United States early in the nineteenth century. Inside the ivory barrel was a ball of thread, the end of which was pulled out through a small hole in the barrel. By the second half of that century, the little ivory containers were intricately carved with landscapes and human figures.[8]

Threading Needles Threading needles were not for sewing, but rather for drawing a length of elastic, tape, or ribbon through a garment. Ribbon needles, for example, made it possible to decorate a lady's garment (or a girl's hat) by drawing a colorful ribbon into and out of slitted segments of the cloth (Fig. 19e).

Yarn Needles These were a special kind of sewing needle, with a large eyehole for the yarn. They were used for sewing together the separate knitted parts of a garment such as a sweater. Although most of them were steel, ivory models have been made.

5 ❦ CONTAINERS: BOXES, CASES, HOLDERS

Containers of every imaginable size, shape, and purpose became a popular way of using ivory. While many were intended as multipurpose receptacles, others were crafted to serve a single and usually obvious function as were candelabra, snuff bottles, and vases. Most were completely enclosed, mainly referred to as boxes or cases; the ones open at the top are usually called holders. Containers whose use can be classified even broadly or remotely under one of the specific-use categories in this book will be considered in that category. The only containers considered in this general category will be those that were intended to fulfill several uses, or are single-purpose containers for which no specific category has been provided.

Ivory collectors should be alert to the fact that some carvings that do not appear to be containers actually have an unexpected compartment inside. A secret compartment is especially unexpected in an ivory product that already has an obvious function, such as a cane. In almost all instances, the compartment is opened by unscrewing an ivory section; the joint is often so difficult to see, we can hardly avoid suspecting that the carver intended that only the owner should know of the compartment. (Practically invisible joints in non-container carvings have often misled collectors into believing that the article was carved from a single piece of ivory.) A caution is in order, however. Collectors who wish to check whether an ivory section can be unscrewed should be extremely careful in applying pressure when twisting the suspect section, lest the section break off.

The earliest ivory containers can be dated as far back as 3000 B.C. in Egypt, and 1200 B.C. in Assyria. By the fourth century A.D., ivory boxes were gaining popularity in Europe. From then on, containers were one of the most frequent articles made of ivory. The multipurpose containers will be considered after the treatment of the containers that had a single intended function.

Animal Cages In Africa small corrals for domesticated animals were sometimes made entirely of medium-size tusks, standing in the ground like wooden posts. Centuries earlier, the Romans and the Chinese used small ivory cages for pets. A Roman reference to an ivory birdcage is found in Martial's *Epigrams,* written during the first century A.D.[1] Laufer cites the Roman writer Statius as telling that his Roman countrymen kept parrots in ivory cages; but Laufer states that not even one of those cages has survived. The Chinese housed pet crickets in small plant gourds that were covered with a fitted ivory top. The insects were kept as "singing" pets, but some enterprising Chinese used them instead for cricket fights, often in matches involving high stakes.[2] A cricket gourd with an ivory cover, from the Chien Lung period (1735–1796), can be seen in the Chinese exhibit at Chicago's Field Museum of Natural History.[3]

The Chinese sometimes housed their pet birds in cages made of ivory.

At the auction sale of the American Art Galleries in 1915, several such cages were exquisitely ornamented with carvings of perches, cups, seed chutes, animal figures, and flowers.[4] At London's Great Exhibition of 1851, an ivory birdcage was on display.[5] In Princeton, New Jersey, the Johnson Whaling Collection had a large birdcage, about 2 feet (60 cm.) high, made of whalebone; it was easily big enough for a parrot.[6] A similar one is at the Mystic Seaport Museum in Mystic, Connecticut.

Calling-Card Cases During China's last dynasty (Ching 1644–1911), calling-card cases made of ivory were elaborately carved; the detailed artwork often portrayed pleasant and relaxed domestic scenes. Before the eighteenth century had ended, the aristocracy of France and England discovered that visiting cards made a greater impression if presented from an attractive container. In fashionable circles in England during the nineteenth century, much importance was placed on carrying these card cases.

The shape and size of ivory cases became standardized as thin rectangular boxes, approximately 4 inches (100 mm.) long, 3 inches (75 mm.) wide, and $\frac{3}{8}$ inch (9 mm.) thick, with room for a small supply of visiting cards. In some, the lid slid off; in others, the lid was hinged on one end. (Besides ivory, there were also card cases made of silver, mother-of-pearl, sandalwood, and other fine substances; affluent people even had a special card case of black ebony for periods of mourning). Often, the artistic embellishment was exquisite. Some cases had a tube on the side containing a dainty pencil for adding a personal note on the back of the card; occasionally the pencil had an ivory cap, or might be made entirely of ivory with a lead point. For the day's reminders to the carrier, some cases had small note cards; some had a thin memorandum slip made of ivory on which a note could be written in pencil and subsequently erased. One kind of card case opened lengthwise like a small book and was sometimes lined with velvet or silk. This type might be spill-proof; that is to say, it could not open unless the pencil was removed from a locking position. Perhaps inevitably, there finally appeared an automatic card case, whereby the press of a button opened the case and presented a single card.[7] In England and British India, large numbers of fine ivory card-cases were made in the nineteenth century.

The name cards themselves had to be appropriately attractive. "Card writing," with the name in flourished style, became a specialized vocational enterprise. The artistic specialists sometimes worked on street corners at small folding desks. Highly embellished visiting cards became miniature works of art.[8]

Figure 28 shows three calling-card cases. One, from China, is entirely of ivory, with carved and painted scenes on front and back. The second case is ornamented with a stag and extremely fine filigree work; it has a note pad, card pocket, and a pencil capped with ivory. The third one, with brass horseshoe and inset turquoise stones, has a satin-lined interior, twelve gilt-edged pockets, and an ivory pencil with a lead point. The origin of the latter two cases is unknown.

Candle Holders An ivory candlestick required a glass or metal insert lest the ivory become burned when the candle burned down to its lower end. Ivory has not been a popular substance for the familiar candelabrum that usually held three candles. In the seventeenth century, however, a large number of ivory candelabra were produced. Nineteenth-century whaler-seamen carved some from walrus ivory.[9] Some seamen carved a candle lantern from pieces of whalebone; one can be seen at the Whaling Museum in New Bedford, Massachusetts.[10] Figure 25 shows a Japanese candlestick, probably late nineteenth century, with a protective insert in the top.

Canes as Containers Although the primary function of canes was of course as an aid in walking, a secondary function was as a container for small articles, sometimes deceptively as a secret container. In such instances, especially when the cane concealed a sword within, the carrier often had no need of the cane as a walking aid. By the time of the Middle Ages, the handle or shaft of some canes had a secret compartment for money, jewelry, valuable documents, and religious relics, or for concealing forbidden items, such as silkworm eggs from China when being smuggled into Europe.[11]

Cases, Caskets, Coffers for Valuables In the classical sense of the term, ivory caskets were wooden boxes (cases or coffers), usually rectangular and of small or medium size, decorated or completely covered on the top and sides with ivory panels. Some were made entirely of ivory. Special ivory containers for miscellaneous valuables date back as early as the fourth century in Europe. In some the panels showed religious themes of Christian significance; other examples had only secular themes. In later centuries, Islamic Spain became famous for its ivory caskets with gold or silver hinges and locks. At the court of Maximilian in Bavaria, the famous German sculptor of ivory Christoph Angermair (early 1600s) made several large chests that used adjacently joined ivory panels throughout.

As distinguished from ordinary ivory boxes, the superlative artistic carving of these cases was the feature that made them desirable as containers for precious and valued articles. (Coin cases will be discussed separately, in Money Cases, p. 99.) Ivory cases can be seen in many of the decorative arts museums in Europe, and also in some of the principal ones in America.

London's Victoria and Albert Museum has tenth- and eleventh-century Moorish caskets, an eleventh-century Byzantine casket, and fourteenth-century French caskets.[12]

Envelope Cases A single reference seen by this writer concerning special ivory cases for envelopes is in a treatise from India; the cases were still being made around 1901, the year the work was written.[13]

Handkerchief Boxes The same (as above) is true concerning special ivory containers for handkerchiefs. Such cases or boxes could of course also be used for holding a variety of other small articles.

Incense Cases and Incense Burners Attractive ivory incense cases were carved in China during the Sung dynasty (A.D. 960–1279); some showed landscapes or animal and bird figures.[14]

Chinese incense burners had traditionally been made of bronze, varying in size from several inches to more than 1 foot (30 cm.) in diameter. The ones made of ivory were of course small and were usually carved with the traditional designs or figures seen in the older bronze models. Barry Eastham's book on Chinese ivory art shows an ivory container which had a carved lid, also of ivory; it looks like a short stubby jar or vase, and was sometimes used for holding flowers.[15]

Money Cases Money cases of ivory date back to the first century A.D. at least. The Roman poet Martial enthuses about luxurious containers, especially those used for gold coins: "Ivory money boxes. To fill these money boxes with anything but yellow money is unfitting; let cheap wood carry silver."[16] Fine cases for coins continued to be fashioned in the centuries following. India was still making ivory money boxes at the beginning of the twentieth century.

In Munich's Bavarian National Museum there is the most elaborate coin display case ever carved. It was made in the Italian style by Angermair some time between 1618 and 1624. The 22-drawer cabinet is said to be about 18 inches (46 cm.) high, not including the base and the equestrian statuette and sitting figures at the top. With the two swinging side-doors open at left and right, the entire ivory cabinet is about 32 inches (81 cm.) wide and high. The front and back of the doors are richly sculpted in relief.[17]

Picture Frames Small- and medium-size ivory picture frames were made in China during the last dynasty (1644–1911) and were popular in India during the nineteenth century. Many claimed to be of ivory, however, are spurious. They were made of Celluloid, often recognizable by the poor imitation of ivory grain, the greater flexibility, and sometimes the implausibly large size.

Pin Boxes Although pin boxes made of ivory are mentioned occasionally, they were probably similar to needle cases, discussed earlier (see p. 94). Or they may have been small simple ivory containers used for any small articles.

Purses and Purse Frames In England, small, firm purse cases were made entirely of ivory, with added metal fittings.[18] Alaskan Eskimos carved walrus-ivory fasteners for ladies' handbags.[19] A pair of ivory purse-frames from China is shown in Figure 22.

Scent Boxes Ivory scent boxes that rested on a table or shelf to add fragrance to the room (as might also be done with incense) should be distinguished from scent bottles for personal grooming. The latter usually had a slender ivory applicator extending down from the stopper (see Perfume Flasks and Perfumers, p. 164). An ivory scent box from eighteenth-century China can be seen at Chicago's Field Museum of Natural History.

Snuffboxes, Snuff Bottles, and Accessories Snuff-taking is the practice of sniffing finely ground tobacco for its stimulating and tickling effect on the nasal membranes. Instead of sniffing the tobacco, some users massaged their gums or the insides of their lips with it. Tobacco smoking articles and accessories are treated as a separate category (see p. 178).

The first description of snuffboxes seems to be in a 1636 publication in Italy, in which ivory examples are also mentioned; but no seventeenth-century ivory snuffboxes have survived. Cheap snuffboxes were made of wood or metal, but royalty and the wealthy soon had exquisite and expensive ones. Gold was the choice material, with silver next in preference; ivory snuffboxes represented a middle ground between the plain and the precious. Many of the finest boxes were embellished with gemstones, portraits, and other ornamentation. Most gentlemen carried a snuffbox and would feel obliged to offer a pinch of snuff to a friend encountered somewhere. In some Western European countries a "snuff etiquette" established the appropriate way of removing snuff from the box and applying it to the nostrils; it prescribed how it should be offered to other people present. After a Great Hundred Years of magnificent snuffboxes—1730 to 1830—the interest in snuff-taking and fine containers for it began to decline, partly due to the newer fashion of cigarette smoking.[20]

Some time after snuff was introduced into China around 1655, carved ivory snuff bottles—but not snuffboxes—began to appear there. A slender ivory or bone spoon was attached to the underside of the cork or stopper, itself usually made of ivory. On the desk or table where snuff was kept, there might also be a tiny ivory snuff saucer about 2 inches (5 cm.) in diameter for convenient use in picking up a pinch of snuff. Only a small amount would be kept that way, because the ground tobacco would soon dry out. In time, ivory snuff bottles were often richly carved in high relief, with landscapes or human figures; poetic inscriptions might be added on the front or on both sides. The snuff saucers, too, were at times embellished.[21]

It is often difficult to distinguish between Chinese and Japanese ivory snuff bottles, because many of the latter were carved in the Chinese style. In her treatise on Chinese snuff bottles, Lilla Perry offers a possible clue: Japanese carvers more often colored or stained their ivories, whereas Chinese carvers usually showed the natural ivory surface. Because of the scarcity of hornbill, snuff bottles made of that substance were rare and expensive. In 1960 hornbill snuff bottles were selling at a minimum of $300; desperate collectors were offering as high as $600. Perry shows an ivory accessory for filling snuff bottles through the narrow opening—a slender combination scoop-and-chute for scooping up a small quantity of snuff and funneling it down into the neck of the bottle.[22] A Chinese set of three ivory accessories (spoon, dish, and funnel) is at the Fowler Museum of Decorative Arts in Los Angeles.

Except in parts of southern Europe, snuff-taking has almost entirely ended. Figure 22 shows an ivory snuffbox and two snuff bottles.

Spill Cases Victorian England provided ivory cases for holding spills— slender wooden strips or twisted paper strips—used for lighting up a lamp or candle.[23] Of course, any long ivory container could hold spills.

Stamp Boxes The single reference seen by this writer concerning special ivory boxes for postage stamps is in the earlier-mentioned publication from India; they were still being made around 1901, the year that work was written.[24]

Vases Ivory vases were found in Egyptian tombs of predynastic times (before 3100 B.C.). Although most ivory vases may have been utilitarian, it seems that the finest of them were often intended purely as decorative pieces. Especially in the case of the perforated type of ivory vase, only dry or artificial flowers may have been kept in them (if they were actually used as containers at all).

Perhaps the most ambitious ivory carvings from China were the large vases carved from a single section of tusk (monobloc carvings). These were sometimes 2 feet (60 cm.) high, worked in relief all around the vase, showing large scenes, intricate designs, and delicate figures. In other Chinese vases, the design sometimes required two or more sections. These were cemented together, one above the other, so skillfully that the joined edges are not visible to the untrained eye. Remarkably, an unfortunate crack in the ivory, perhaps discovered after carving had begun, was sometimes disguised by making it serve in the design, for example, as a fold in drapery or as a branch of a spray of blossoms.[25]

Probably the finest examples of vases as craftsmanship and pure art were the two-piece marvels made by Russia's great carver N. S. Vereschagin (1770–ca. 1814). Mostly commissioned by the Czar, they looked like large pierced ivory eggs resting on decoratively carved ivory bases. Stunning examples of such so-called Palace Art of the eighteenth and nineteenth centuries are now in the art museums in Leningrad and Moscow.[26]

A vase from India illustrates a charming royal practice from earlier times—the Tulabharam ceremony. We see the new maharajah being weighed on one side of a balance scale, against a huge stack of gold coins on the other side. (These coins were specially minted for the occasion.) After the weighing, noblesse oblige required that the entire pile of gold coins be distributed among the officiating priests and Brahmans.[27]

A publication of almost a century ago tells that nineteenth-century India had "flower stands" (presumably for dried or artificial flowers), made of ebony wood with ornamental ivory inlays.[28]

Figure 24 shows a Chinese vase, probably mid-nineteenth century, attractively carved and painted with scenes on both sides and with a pair of hanging ivory rings.

Water Vessels Literature on ivory products rarely mentions the secular use of ivory for water vessels. (In this book, church use is considered in a separate section; see Vessels for Holy Water or Wine, p. 176.)

Ancient production of water containers made of ivory is suggested by finds of Chinese pail-shaped vessels (from the first dynasty, Shang, 1500–1027 B.C.) made from the widest section of elephant tusk, and decorated with figures of cicadas and some kind of monster.[29] In times past, magnificent ivory pitchers were carved. Two of them—20 inches (50 cm.) high and luxuriously carved—were in the George A. Hearn Collection at the beginning of the twentieth century.[30] From Ceylon, a centuries-old water-dipper with an ivory handle is now in London's Victoria and Albert Museum.[31]

*   *   *

A very large number of ivory containers of the past appear to have been intended as general purpose receptacles. Of course, many kinds of special purpose containers could be used for holding any small articles.

From early dynastic periods, Egyptian tombs had ivory strips that apparently had once covered wooden boxes stored there; the ivory showed geometric designs. The tomb of King Tutankhamen (fourteenth century B.C.) had large boxes veneered with carved and painted ivory. A large wooden chest was veneered with ivory on the top and sides; the ivory is carved in relief and is colored, showing different scenes and floral elements. This chest is considered an artistic masterpiece of that period.[32] At the site of ancient Megiddo in Palestine, an ivory box of Phoenician style dating back to the period 1350–1150 B.C. was found.

An occasional and somewhat curious kind of carved container in nineteenth-century India was a shallow basket made entirely of long fine filaments of ivory; it required almost three months of full-time labor, and its intended use is now uncertain.[33] Even earlier, John Barrow, visiting China in 1792, described baskets (and hats) made of interwoven ivory shavings and quill pieces[34] (see Bread Baskets, p. 114).

Figures 21 and 23 show various containers of uncertain or miscellaneous use. The seven pictorially etched mother-of-pearl discs, shown in Figure 21 with their pair of finely carved Chinese cases, may have been used as game markers or as coasters for tiny wine glasses.

6 ❖ COSTUME ARTICLES AND ACCESSORIES

Costume articles of ivory are among the oldest of uses for this attractive substance. In Egypt, ivory for personal adornment dates back at least five thousand years. Although most ivory items of adornment sufficed with that valued material alone, some pieces were further embellished by the addition of substances of greater value, such as gold, silver, or precious gemstones. As will be noted, even the accessories for costume articles were sometimes made of ivory.

Bracelets and Bangles Narrowly defined, "bracelet" refers only to the band worn on the wrist, whereas "bangle" (from the Hindi) refers more broadly to any band worn on the arm, similar to the term "armlets." Ivory examples of this kind of jewelry were found in Egyptian tombs of predynastic times (before 3100 B.C.). The ancient Indian town of Pali, on the old trade route between Bombay and Delhi, turned ivory bangles for several centuries. Entire streets of craftsmen made only this article, in sets of graduated sizes, so as to cover a woman's arm from shoulder down to wrist. They were cut from the tusk as it too decreases from base to point.[1]

In some districts of the Punjab in northwest India, a girl could not wear bangles until her wedding; then, at the wedding ceremony, she had to wear a bangle, preferably presented by a maternal uncle.[2] No doubt more ivory bangles were presented too. Particularly in the eighteenth century, India financed a substantial ivory trade from Portuguese East Africa. In exchange for large quantities of Indian cloth and beads, the preferred "white ivory" (African, rather than Asian) was obtained for making the numerous bangles presented at wedding ceremonies.[3] Figure 27 shows two solid ivory bracelets, one from Southeast Asia, the other from Africa; an ivory bracelet-and-earring set and a similar set including a matching necklace are shown in Figures 4 and 26 respectively.

Ancient Egyptian graves are said to have yielded ivory anklets. Although we may wonder how the discoverer could know that an article had been intended for the ankle (unless it was found being worn that way on bodily remains), it seems likely that some ivory anklets have been fabricated in times and places where fashion made them desired.

Brooches Ivory brooches appeared in England as early as the twelfth century.[4] It is likely that decorative ivory pins were worn in ancient times. Tastes in costume jewelry of this kind have changed over periods of time. For example, Victorian jewelry for women of fashion began as small simple items—brooches, necklaces, wrist bracelets, and so on. But by the mid-1860s, only a few decades later, such ivory jewelry had become larger and more ornate, carved intricately, and often inlaid with gemstones; it was a style that was later considered to be ostentatious. Before the nineteenth century ended, fashionable ladies began to prefer delicate filigree in their costume jewelry.[5]

Figure 38(a) shows a Chinese or Japanese brooch with the character for "happiness" (in Chinese, *fu;* in Japanese, *fuku*) at its center, symbolizing "good fortune," and similar to the English "Good luck" wish.

Brow-Bands Ivory brow-bands made by Alaskan Eskimos are mentioned occasionally; the present writer has not seen them, nor even a description of their appearance.

Buckles An elaborately carved ivory buckle from China's Chou dynasty (1027–256 B.C.) is at the Fogg Museum of Art at Harvard University.[6] Alaskan Eskimos have made such belt buckles from walrus ivory; those articles for the commercial market are incised quite attractively.

Busks Starting in the 1700s, a thin shaft of whalebone or baleen 8 to 12 inches (20–30 cm.) long, 1 to 1½ inches (2.5–3.8 cm.) wide, was inserted into an open vertical pocket in the front of a fashionable lady's corset or bodice. It served as the mainstay for firming up the front of the garment. Many were decorated with pictorial representations, although the decoration could not be seen when the busk was in use. Examples of busks can be viewed at the Nantucket Historical Association and at New Bedford's Whaling Museum, both in Massachusetts.[7] Figure 36(a) shows a busk bearing a small sketch of a ship; probably more art work had been intended.

Buttons Although buttons have been used for thousands of years now, they became especially popular during the 1200s, when fitted clothes began to replace robes and loose garments secured with cords, belts, or pins. The earliest buttons were made of simple natural substances, primarily wood and bone, and had pierced holes.

The use of ivory for buttons started some centuries ago. China and Japan were apparently the main sources of attractive hand-carved ivory buttons; the painted or carved themes were predominantly flowers, dragons, and Oriental scenes. On the back of some old Chinese carved ivory buttons we can see the beginning of the clever practice of cutting a "tunnel hole," with both mouths on the back side, so that the artistic portrayal at the front would not be spoiled by pierced holes; the front might show several insects made of bits of gemstones in rich colors. Highly artistic Italian ivory buttons typically showed flowers or birds, often in deep carving and exquisite detail. Alaskan Eskimos carved simple buttons out of ivory and bone. Figure 38 shows several types of clothes buttons made of ivory, including collar buttons (or studs) for fastening a detachable collar to the neckband of a shirt, or for fastening together the ends of a neckband.

By the middle of the nineteenth century, the race had started for the development of preferred substances for button manufacture, substances that would be cheap, readily available, and suitable for the required manufacturing operations. The use of vegetable ivory for buttons and similar articles was discussed in Part One (see p. 38). In 1855 in England, Alexander Parkes concocted an early composition material called Parkesine that was intended to replace the use of india rubber. Soon the invention of Celluloid and other plastics began to replace the use of vegetable ivory.[8]

Buttonhooks Almost as soon as shoes, and then gloves, became fitted with buttons for fastening the left and right edges, it became necessary to develop a hook for reaching through the small, tight buttonhole and pulling the button back through it. The shank was of slender metal construction, terminating in the familiar question-mark form of hook. Ivory was among the more valued substances used for the handles. Unlike shoe buttonhooks (Fig. 36b), glove buttonhooks were very short (Fig. 36d) because it was not necessary to reach down low in order to use them. Buttonhooks for gloves, collars, starched shirts, and spats (instep covers) sometimes had a double-prong form, or a metal loop in the shape of a diamond or a narrow ellipse, instead of the typical question-mark hook. The point of the diamond-shaped hook was also used for prying open any tightly starched buttonholes on a shirt or collar. Figure 36(c) shows a buttonhook like this; the notch in the handle is unusual and may have been intended for firmly holding up spats buttons while pressing the buttonholes around them. As with the handles of other ivory products, the ones cut by machine were the most common, while the hand-carved ones were expensive and are therefore relatively rare.[9]

Ultimately, buttonhooks became obsolete because the above kinds of button arrangements were no longer used. Detachable collars and instep spats disappeared entirely. Shoe buttons were replaced by shoelaces, zippers, and elastic form-fitting shoes. Glove buttons were replaced by snap fasteners and by gloves without any fastener device. Even starched shirts, with their tightly closed buttonholes, fell out of fashion.

Clothes Brushes Large handles and backings, as on clothes brushes and hairbrushes, were often made of simulated ivory, starting with the invention of Celluloid. More expensive clothes brushes have gained from the clean light color of real ivory. Figure 36(e) shows a clothes brush with a monogrammed ivory back; it is part of a set with a matching hairbrush. (See also Hat Brushes, p. 107).

Clothespins In a period when most wives washed the family's clothes and then hung the wet wash up to dry, some whaler seamen's wives used clothespins made of whalebone.[10] Some of these can be seen at the Mystic Seaport Museum in Mystic, Connecticut.

Cuff Links Cuff links were among the few ivory costume articles made exclusively for men.

Earrings Ivory earrings were especially popular when they were part of a matching set, in combination with a bracelet or a necklace, or both (see Figs. 4, 26).

Glove Boxes The only reference to ivory boxes for gloves seen by this writer is in the treatise from India mentioned earlier. They were still being made there around 1901, the year the work was written.[11]

Glove-Powder Boxes Here, too, a single reference mentions the making of ivory containers for glove powder.[12] Presumably, the powder, applied to the hands or inside the gloves, facilitated the putting on and removal of tight gloves.

Glove Stretchers When ladies' leather gloves became wet in the rain, or when cloth gloves were washed, it became necessary to restore the shape and enlarge the shrunken size, especially of the glove fingers. Usually made of wood or metal, some of the choice models were of ivory. The stretcher's adjacent pair of tongs were inserted into one glove finger at a time. By squeezing together the two sides of the hinged handle, the jawlike tongs opened, thus stretching the inside of the finger (Fig. 37).

Hats or Caps John Barrow, visiting China in 1792, described hats (and baskets) made of interwoven ivory shavings and quill pieces.[13] As a substitute substance, baleen (from baleen whales) has been used by nineteenth-century American whaler-seamen for making pieced caps or hats.

Hat Brushes Hat brushes looked like large toothbrushes or very small clothes brushes. When the head (holding the brush elements) was made of ivory, the handle might be of sterling silver in a decorative style (Fig. 37).

Hat Stands Chinese carvers made short ivory hat-stands for holding skullcaps, those small, close-fitting caps without a brim.

Hatpins and Hatpin Stands Especially during the period from 1880 until the beginning of World War I in 1914, hatpins were popularly used for securing ladies' large hats to a sizable mass of hair. The top of the metal pin was made in a variety of materials, including ivory (Fig. 37). At the beginning of the twentieth century, it was reported that the special stands for holding a lady's hatpins were sometimes made of ivory.[14]

Horse Trappings For ornamenting both horse and rider, we may note a pair of artistically carved ivory stirrups crafted in the last Chinese dynasty (1644–1911). Now in a private collection, they may have been used during ceremonial appearances of members of the royal family.[15] They are possibly the only ivory stirrups in existence.

Animal costuming, too, has used ivory for embellishment. During the early Roman Empire, horse saddles were sometimes decorated with inlays of ivory pieces. In twelfth- and thirteenth-century England, the better saddles and other horse trappings had such ivory inlays.[16] In Paris, the Louvre museum has a carved ivory saddle-bow (upper front part of a ceremonial saddle) from the thirteenth century; similarly, it has a carved ivory cantle (upper rear part of a saddle) that was made in the early fourteenth century for King Frederick of Sicily.[17]

Jewelry Boxes and Stands Special ivory cases for other valuable articles are discussed under the general category of Containers (see Canes as Containers, p. 98; Cases, Caskets, and Coffers, p. 98; and Money Cases, p. 99). The tomb of King Tutankhamen (fourteenth century B.C.) included a large rectangular jewelry chest; the sides and the arched lid were covered with ivory veneer and had inlaid panels of ebony and ivory marquetry.[18] In China, ivory cases for jewelry date back at least to the Tang dynasty (A.D. 618–906). They were being carved in England during the twelfth and thirteenth centuries. In the early 1400s, the northern Italy workshop of the Embriachi family produced exquisite jewelry cases which became famous examples of ivory workmanship. Figure 40 shows a French jewelry case (*ca.* 1840) made of ivory and brass; the interior is lined. Islamic jewelry boxes of wood with tiny ivory inlays were of extremely intricate design. Nineteenth-century American whaler-seamen sometimes carved a "jewelry tree" from whale-tooth ivory, with individual branches for holding the separate pieces of jewelry.[19]

Labrets Similar to other facial jewelry (earrings in pierced ears, or nose rings in the pierced nasal septum), Alaskan Eskimos once wore ivory labrets on their face. The small ivory pegs were worn in shallow holes pierced through the skin at the left and right sides of the lips.[20]

Mourning Jewelry Special, less ostentatious jewelry used during (and signaling) a time of mourning appeared in England in the nineteenth century, reaching a height of popularity at the death of Prince Albert, when Queen Victoria began a prolonged period of mourning for her abruptly and prematurely deceased husband. Fashionable ladies indicated their distress when a loved one passed away by removing their glamorous jewelry and substituting the somber kind that thoughtful and sympathetic artisans were ready to provide. Appropriately, the mourning jewelry—rings, brooches, lockets—were usually made of jet, the deep-black coallike substance that takes a glossy polish.[21] Some were made of ivory, with a black trim. The mourning lockets often contained a memento, such as a tiny picture of the departed family member, or a lock of hair of a deceased child.

Neckerchief Slides Nineteenth-century American seamen, as well as American Civil War soldiers, sailors, and prisoners-of-war, carved neckerchief slides from sections of bone.[22] These were for holding together the corners of a neckerchief and looked like small napkin rings. It is not certain whether any were carved from ivory.

Necklaces Ivory necklaces were found in Egyptian predynastic tombs, dating earlier than 3100 B.C. In India, ivory beads have been found that may date back to 2700 B.C.[23] It appears that ivory necklaces have been carved worldwide. One of the few delicate types of early Alaskan Eskimo carving was filigree ivory necklaces, intended more as works of art than as costume articles.[24] Their twentieth-century ivory productions, however, have included large numbers of matched sets of necklaces and earrings made for sale as costume items. The individual beads of ivory necklaces have been carved in the form of flowers, animals, geometrical designs, and so on (see Fig. 26).

Rings As with necklaces, discussed above, the earliest ivory rings were found in Egyptian predynastic tombs. From China, ivory rings are thought to date back to at least the Chou dynasty (1027–256 B.C.). They are mentioned in ancient Roman literature; the poet Statius (first century A.D.) tells of "the ivories and the gems worthy to adorn a finger."[25] The earlier section on walrus ivory in Part One (see p. 31) related that Scandinavian warriors once believed that ivory rings could magically ward off an attack of body cramps. In Germany, Stephan Zick (1639–1715) carved ivory curiosities, including some sets of trinity rings—three intertwined and inseparable rings, carved from a single small piece of ivory (one set is in London's British Museum). Figure 38(f) is an ivory ring with the side designs carved in deep relief; the front surface appears to have been prepared for the monogram or design that was never carved.

Sandals Ivory sandals have been mentioned in some writings, but this writer knows of none that may have survived.

Shawl Clips These were apparently used for fastening the ends of a lady's shawl. An example of a shawl clip made of whale ivory can be seen at the Mystic Seaport Museum.

Shoehorns Both plain and decorative ivory shoehorns have been carved.

One from Japan with head of a girl is shown in Figure 37. As substitute substances, baleen (from baleen whales) and whalebone were used too. Pairs of boot hooks made of whalebone were used as an aid in pulling on heavy leather boots. The user inserted one hook into each of the two straps at the top of the boot, and pulled hard.

Slipper Clips   The upper section of a plain pair of slippers could once be embellished by attaching an attractive pair of small plaques, sometimes made of ivory. A pair with carved floral design is shown in Figure 38(g).

Snow Goggles   For protection against the intense glare of sunlight reflected from snow and ice, Eskimos made snow goggles from a single thin strip of walrus ivory, with two narrow slits that admitted only enough light for vision.[26]

Walking Sticks as Costume Articles   Probably the earliest example of a ceremonial walking stick embellished with a carved ivory handle was the one found in the tomb of the Egyptian King Tutankhamen, from the fourteenth century B.C. It bears the carved figures of the king's two enemies, shown as captives bound to the stick. The "northern enemy" is an Asiatic representation, bearded, with the face carved of ivory. The "southern enemy" is an African, with face carved of ebony.[27]

After the middle of the fifteenth century A.D., walking sticks began to serve a more elegant purpose, becoming articles of fashion rather than simply practical aids. In the eighteenth century, a cane etiquette was developed that governed the carrier's conduct with walking sticks. Ivory was among the fine materials used for these costume additions. As recently as the times of Edwardian England, it was not unusual for a gentleman to own a set of walking sticks, each one suitable for a different kind of occasion. The ones with inner compartments were often carried as showpieces, except perhaps the secret container canes mentioned earlier (see Canes as Containers, p. 98). Special canes were designed to hold or convert into an unexpected article, such as a coin dispenser, fan, fishing rod, flashlight, liquor flask, perfume bottle, ruler, smoking pipe, telescope, or umbrella. Some of these had an ivory handle (Fig. 35).

The use of ivory (or any valued material), especially any decoratively carved ivory, had helped to establish the fashionable function of walking sticks. Perhaps this non-practical use is illustrated most starkly by the canes made for boys in the nineteenth century, imitative of the walking sticks carried by their stylish fathers; although not made in large numbers, these canes sometimes had ivory handles.[28]

Washtub Clothes-Fork and Clothes-Paddle As accessories for the cleaning of costume wear, we can consider two products made of an ivory substitute. For stirring and lifting the plainer clothes being washed in hot water, the wives of whaler seamen had large forks and paddles made of whalebone. A lift paddle is at the Mystic Seaport Museum; this and a stir fork are shown in Flayderman's book on scrimshaw.[29] The two kinds of aid were used interchangeably for either stirring or lifting.

7 ❋ EDUCATIONAL ARTICLES

Devices used in discovering or processing information (such as calculators) are discussed under the category of Scientific Instruments and Aids (p. 177). The present grouping is limited to ivory products intended to display the already known. When we consider the large number of applications of ivory in other categories of products, it may seem surprising that ivory appears to have had only a small number of uses for educational purposes. As in the instance of other categories, educational ivories were usually attractively carved.

Alphabet Sets and Tablets In nineteenth-century India and China, a simple educational aid for children consisted of English alphabet letters; the better quality sets were made of ivory. English alphabets were presumably intended for foreign export, but it is likely that British personnel in India bought some. Ivory alphabets are rarely seen today; they are a typical example of a set that is sooner or later thrown away when some of the pieces are lost.

During the same century, American whaler-seamen used whalebone to carve alphabet sets, as well as alphabet tablets that showed all the printed letters on one side.[1]

Figure 44 shows an attractively designed alphabet set from China. Inclusion of an ampersand, where "and" seems irrelevant, is odd; could this set have had a different purpose, perhaps as a game or puzzle?

Anatomical Models In Nuremberg, Germany, Stephan Zick (1639–1715) carved ivory models of the eye and ear, including some of the internal mechanisms used in seeing and hearing (as understood at that time). Some of the models are in the German National Museum in Nuremberg (Fig. 39), although a few in that collection have been damaged by the bombings of World War II.

Another display model is a small ivory carving of a complete human skeleton, mounted in a standing position in the Museum of Decorative Arts in Paris.[2] The National Museum in Copenhagen has a large and rare ivory instructional model of a skeleton 28½ inches (73 cm.) high, carved by Niels Gyntelberg in the seventeenth century.[3] The Anatomical Institute in Basel, Switzerland, has an ivory instructional model of a skull, 5.3 by 4.1 inches (13.4×10.3 cm.).[4]

Stephan Zick also carved small ivory models of a female figure and a male figure, these being said to be about 7 inches (18 cm.) high. The front of the trunk can be removed to reveal the internal organs. Those models are reported to be in the History of Medicine Museum in Copenhagen. A similar model of a female figure is by an unknown ivory sculptor; it is 6.4 inches (16.2 cm.) high, carved around 1740 in Germany, and is now in the Kestner Museum in Hanover, Germany.[5] In Nuremberg, a magazine in 1803 listed for sale an ivory model 8 inches (20 cm.) high of a pregnant woman. The front of the abdomen could be opened.[6] Did it contain an ivory fetus? The present writer's written attempts to locate the European reporter of that information, published in 1961, were finally successful. A reply was received to the effect that the model (and other similar ones) did indeed contain a miniature ivory fetus.[7]

Architectural Models The English carver Richard Lucas (1800–1883) made two scale models from ivory to show what the Parthenon in Athens looked like before and after it was blown up in 1687 when a Turkish ammunition magazine accidentally detonated. This pair of before-and-after demonstrational models are in the British Museum.[8]

Maps Old ivory billiard balls were sometimes converted into world globes by painting a world map on the surface, presumably after sandpapering or otherwise removing any markings on the old surface. Both terrestrial and celestial globes painted on ivory balls were made in Western Europe in the seventeenth and eighteenth centuries.[9] Even the geographic ones are seldom seen anymore; an ivory celestial globe (showing some constellations of stars) must be a true rarity. Figure 44 shows a terrestrial globe produced by scrimshaw work on an ivory pool ball, apparently done in 1855 after a long voyage of the *Whaler Hawk* (or a whaling ship named *Hawk*). The geographic sizes and shapes are somewhat erroneous for a few of the drawn continents, but a redeeming feature is the fact that the route is shown for the entire world voyage. Starting at the east coast of the United States, the ship sailed southeastward around Africa, and continued on through the Indian Ocean, Indonesia, New Zealand, and Australia; she then rounded the southern tips of Africa and South America, sailed up to the west coast of the United States, turned back to follow a course around the entire coast of South America, and, at last, sailed home to the east coast of the United States. Such a commercial voyage, including stops, probably required at least two years.

Pointers for Blackboards and Charts Nineteenth-century American whaler-seamen carved some long pointers from whalebone.[10] Although whalebone was an inexpensive substance, collectors now would have great difficulty obtaining one of those pointers.

8 ❁ FOOD-RELATED ARTICLES

Directly or indirectly, ivory articles have had a very large number of applications related to foods; they have been used in either the preparation, the presentation, or the eating of foods. And, as will be noted, ivory has even been used as a food itself. Some articles were made very plainly, with utility the only consideration. Others have been richly carved; in some the artistic embellishment is lavish.

Apple Corers Apple corers made of ivory have been carved by nineteenth-century American whaler-seamen.[1] Apparently, few have survived.

Choppers A whaler-seaman carved a solid whalebone vegetable-chopper.[2] Alaskan Eskimos made their *ulu* cutter with a walrus-ivory handle across the top of the cutting blade.

Chopping Blocks Occasionally, the literature on ivory mentions that large flat sections of ivory tusks have been used as chopping blocks; most of these were made and used in Africa.

Chopsticks and Chopstick Holders Although the Japanese writer Sen-rei Soma, mentioned in Part One (see p. 18), spoke of Chinese ivory chopsticks in the Han dynasty (206 B.C.–A.D. 221), later investigation in China discovered that ivory chopsticks were being used even earlier, during the Chou dynasty (1027–256 B.C.).[3] Until the decline of the use of ivory for chopsticks, Shanghai remained a major producer of them in China. Large numbers of them were also made in Japan and other countries of the Far East. Figure 31 shows an unusual pair of ivory chopsticks from China—only when brought side by side can one see the complete pictures. Chopstick holders made of ivory have come from Japan, and perhaps from China too.

Coasters Coasters, for protecting a table's surface, were sometimes made of ivory. The only ivory coasters known to this writer have come from Japan, but it would seem likely that they have been produced by other countries as well.

Cocktail Picks Ivory cocktail picks, speared into an olive or a maraschino cherry, are no more effective than the usual wooden or plastic ones, but they do add the touch of ivory's elegance, even if only for a few fleeting moments. They may also have been used for appetizer snacks (Fig. 31).

Cookie Cutters and Molds Nineteenth-century American seamen carved these from whale-tooth ivory.[4] They have now become museum pieces, beyond collectors' reach.

Dishes, Bowls, Bread Baskets Small ivory dishes, pieced dishes of larger size, and bowls have been carved since ancient times. Ivory bowls for foods were among the variety of Assyrian ivory artifacts (1200–600 B.C.) found during the excavations at Nimrud in Iraq.[5] A small walrus-ivory dish from China was collected in 1923 by the Marshall Field Expedition for Chicago's Field Museum of Natural History.[6] Small ivory bowls are said to have been made in Japan.

In his book *The Deipnosophists,* Athenaeus, a third-century Greek writer in Egypt, tells of bread baskets that were "made of strips of ivory curiously plaited together."[7]

Drinking Vessels Secular vessels for water are discussed in an earlier category (see Water Vessels, p. 102, under Containers: Boxes, Cases, Holders); church use for water and wine is considered later in Vessels for Holy Water or Wine, page 176, under Religious Articles.

Cups, goblets, and tankards of ivory, used especially for wine or beer, were prominent from the fourteenth century to the beginning of the eighteenth century. In the seventeenth century, ivory beermugs—especially the richly carved tankards with a hinged lid also made of ivory—had become favored possessions. Lucas Faydherbe, seventeenth-century Flemish pupil of Rubens, carved some remarkable tankards. On the ivory drum of one tankard attributed to him there is shown a "mad dance around of nymphs, satyrs, and little bacchanalians." It is now in London's Victoria and Albert Museum.[8] Finely carved ivory tankards can be seen in the museums at Munich, Dresden, and Vienna, as well as in the British Museum.[9] In more recent times, nineteenth-century American whaler-seamen carved cups and also elaborate chalices from whale-tooth ivory and from whalebone.[10]

The use of ivory as a liquid container has, however, two main disadvantages: first, the contact surface will gradually dissolve minute particles of ivory into the liquid if kept in the container for extended periods of time; and second, colored liquids will stain the inside surface. Both of these limitations are avoided if the inside is lined with glass or similarly resistant material.

Eating Utensils Ivory chopsticks are discussed in an earlier section (see p. 113). Egyptian predynastic tombs (before 3100 B.C.) yielded ivory spoons and ivory-handled table-knives; the knife handles showed carved scenes of hunting and war. China's first dynasty (Shang, 1500–1027 B.C.) produced knives with ivory handles. Ivory spoons were found in the Assyrian excavations of the period 1200 to 600 B.C. in Nimrud, Iraq. Ivory utensils for use in eating continued to be made from ancient to recent times. Nineteenth-century American seamen carved solid spoons, forks, and knives from whale-tooth ivory and whalebone (as eating utensils or food-preparation articles).[11]

Solid ivory spoons with elaborately carved handles made in seventeenth-century France are in the Cluny Museum at Paris.[12] A specialized treatise on exquisitely crafted eating utensils shows some superb ivory examples from the late 1300s to the end of the eighteenth century, the century in which the demand for ivory handles for knives reached a peak. Shown is a set of twelve table knives from France; the carved handles portray the full figures of twelve saints. From the Victoria and Albert Museum in London, there is shown a set of fourteen table knives made in England in 1607; the ivory handles show the full figures of English monarchs from Henry I to James I. Also illustrated are excellent examples by Dutch, Flemish, German, and Italian carvers.[13]

In addition to the standard utensils, many specialized kinds were made of ivory. Figure 31 shows items that include a pickle fork and two small spoons; although the larger of the two spoons may be for a baby, the other one (with painted figure, and probably Chinese in origin) is so small that it seems to have been intended for use in adding salt or spices when preparing foods, rather than as an eating utensil. (More unusual possibilities include use as a snuff or opium spoon.) Figure 42 includes a cake knife with a plain ivory or bone handle and a decorative silver-plated blade.

Flour Tryers This device was a flat paddle which was used to pick up a portion of flour from the store's flour barrel for quick inspection in order to detect the possible presence of insects before that portion of flour was deposited into a paper bag. This was of course in the days before prepacked foods. Ivory models were made by the Pratt-Read Company in Connecticut.

Part Two : Uses of Ivory

Ivory Food The constituents of ivory substance are the same as those in some of the foods that animals and humans eat and live by. In some ivory-carving countries, residual ivory dust was fed to cows; it was claimed that this increased the production of milk.[14] In China it was made into a food compound called ivory jelly.[15] The same is said to have been done in Wales. At the end of the nineteenth century, it was reported that "Confectioners and chefs make use of ivory dust."[16] Aside from any basic considerations of nutritional value, it has been said that ivory jelly was being used "to amuse the novelty-seeking palates of the wealthy."[17]

Jar Lids Small ivory lids for jars (probably food containers) from the period 1350 to 1150 B.C. were among the artifacts excavated at the ancient city of Megiddo in Palestine.

Knife Rests Many of the well-equipped homes of the late 1800s owned a set of knife rests. This extra touch of table elegance was sometimes made of carved ivory, with imaginative representations of birds, children, or mythic subjects.[18] A simple but attractive knife rest is shown in Figure 31.

Ladles and Scoops Ladles, used for serving a portion of soup, stew, or similar food, were sometimes made of ivory. Nineteenth-century American seamen carved some sugar scoops and flour scoops from whalebone. A flour scoop can be seen at the Mystic Seaport Museum in Mystic, Connecticut.[19]

Marrow Spoons The functional part of a marrow spoon resembled a small scoop and was used to remove the marrow from bones split open lengthwise. Some have been made with a handle of ivory (Fig. 42).

Menu Cards In an article about ivory-inlaid ebony carvings, a writer in India listed "menu cards" as one such product.[20] Presumably this was a card holder, similar to the picture frames listed in the same article.

Napkin Rings If utilitarian products had to be rated according to how useful or how much needed they actually are, it is likely that napkin rings might be near the bottom of the list. But, like a bowl of fresh flowers on a dining table, attractive holders for napkins helped to introduce an elegant table setting. Napkin rings made of ivory were often artistically embellished (Fig. 31).

Orange Peelers The only orange peelers made of ivory known to this writer are the ones that were made in Japan. A decorative example is shown in Figure 31.

Pastry Cutters, Jagging Wheels, Pie Crimpers As carved by nineteenth-century American seamen, the early ivory models consisted of a simple handle and a thin rotating wheel disk that had a series of serrations on the rim. The serrated disk could cut flat pastry into smaller sections and could cut a sheet of dough into narrow strips (with fancy serrated edges) that were then laid criss-cross as the partial crust on a pie. It could also trim, crimp, and seal together the top and bottom edges of a pastry shell for the crust of a pie. Some had also a pointer or a fork (usually at the beginning of the handle) for making decorative point marks and for perforating the top layer of dough so that the steam could escape.

Later, however, the seamen carved these articles more elaborately than probably any other utilitarian product that engaged their skill. Extra wheels (even as many as six) were sometimes incorporated into the design, all the wheels except one being superfluous because they were almost identical. The seamen, including the captain, vied with one another in carving intricate and beautiful examples. Annual exhibitions were held in the New Bedford Town Hall in Massachusetts, and prizes were awarded. The Old Dartmouth Historical Society in that town has over two hundred such items.[21]

Pastry-Decorator Syringes A small number of these tubular devices were carved from whalebone by whaler seamen.[22]

Pestle-and-Mortar Sets In Africa these food mashers made of ivory date back to the 1800s, perhaps even earlier; one use was for mashing bananas before cooking them. More generally, the combination set was used for grinding grains, nuts, and other seeds. For pulverizing dried spices and certain dry foods, such sets were carved from whale-tooth ivory by whaler seamen already mentioned.[23]

Plate-Scraper Spatulas Seamen carved these from whalebone to make the dishwashing chore easier.

Rolling Pins Similarly, whaler seamen carved wooden rolling pins that had handles of ivory made from whale teeth or walrus tusk; sometimes the entire rolling pin was made from a single piece of whalebone.[24] Examples of these are at the Mystic Seaport Museum.

Salad Mixing Fork-and-Spoon Sets For the export or tourist trade,

[118] Part Two : Uses of Ivory

African ivory carvers made attractive salad mixing sets, some carved in relief with animal or human figures. Figure 42 shows a set made of solid ivory, with carvings on the handles.

Salad Tongs   Salad tongs made of bone were carved by French prisoners-of-war in England from 1795 to 1815.[25]

Salt Cellars and Salt-and-Pepper Shaker Sets   The early Afro-Portuguese ivory trade commissioned the native carvers to make small pedestal cups for use as salt cellars in Europe.[26] Cellars and shaker sets of ivory have been carved in various countries. Using whale-tooth ivory, American seamen carved shakers for salt, pepper, and powdered sugar. Figure 31 shows a salt-and-pepper shaker set embellished with painted and gemstone floral representations.

Strainer Spoons   Although it would be easy to make an ivory strainer-spoon by simply drilling some holes through the bowl end of a spoon, they are rarely mentioned in any writings about ivory utensils for the kitchen. The present writer has seen only one illustration of an apparent example of such a product.[27]

Tea Caddies, Tea Scoops, Teapots   The 1700s furnished highly decorative containers for dried tea leaves; some tea caddies consist of a wooden box covered with a veneer of ivory.[28] Ivory scoops for dried tea leaves have been made in Japan. Also, small teapots have been made from ivory. The disadvantage of using ivory for containing liquid is discussed under Drinking Vessels, page 114; the resulting problems would be even more serious with very hot liquids.

Trays   Only a small number of ivory trays have survived; probably very few were made. Constructed of separate ivory pieces, they were designed so cleverly that it is often quite difficult to see the separate sections. Some were made in China during the last dynasty (1644–1911).[29]

9 ❋ FURNITURE AND HOME FURNISHINGS

The use of ivory for home furniture, furnishings, and decoration goes back more than two thousand years. Some articles were made entirely of ivory, but most of them combined ivory with other materials. Because large-scale use of ivory was quite expensive, the use of small inlays for ornamentation was a popular feature. In India, for example, ivory inlays decorated cabinets, tables, desks, chairs, picture frames, and many other items.

Beds A very early reference to "ivory beds" (presumably just the four bedposts) is their mention, as dismal luxury, in the Bible (Amos 6:4). Even earlier, ivory inlays ornamented the wooden beds found in Egyptian tombs of the period 2850 to 2600 B.C. From ancient Syria, beds decorated with ivory date back to the fourteenth century B.C. Already mentioned in the historical review in Part One was the lament of a certain Roman poet concerning the extravagance of using ivory for bedposts (see p. 20).

Cabinets Specific purpose and general purpose cabinets have been decorated with ivory inlays. In India fine ebony wood was often used for the basic structure upon which the ivory ornamentation was imposed. Probably used for cabinets, small ivory knobs were discovered during the excavations of Assyrian ivories (1200–600 B.C.) at Nimrud, Iraq. Figure 66 includes an ivory article that may have been a cabinet knob (see p. 203).

Chairs The British Museum has ivory-inlaid chairs from ancient Egypt dating about 1500 B.C. A thousand years later, a biblical account tells of "benches of ivory" (Ezekiel 27:6). Since ancient times, fine chairs have often had decorative inlays of ivory; some splendid seventeenth-century Italian examples are in London's Victoria and Albert Museum. The museum also possesses very rare items, made in the eighteenth century—chairs that are made entirely of ivory, except for the padded seats.

Couches From Syria, couches decorated with ivory date back to the fourteenth century B.C. At the Victoria and Albert Museum in London we can see an ivory-inlaid couch made in Italy about three thousand years later. As recently as the nineteenth century, the Maharajah of Benares in India had an elaborately carved ivory couch.[1]

Footrests and Headrests Occasionally, ivory footrests are mentioned; but the present writer is unaware of any surviving examples. Probably, headrests of ivory (among other materials) were made more frequently, though few examples can still be seen. Two of them were found in the tomb of King Tutankhamen in Egypt (fourteenth century B.C.).[2] The simpler one is a folding type of headrest. The other one is a splendid carving; the semicircular upward-curving headrest is held up by an Egyptian on his knees, between two resting lions. A recent treatise on King Tutankhamen's tomb identifies the Egyptian as Shu, god of the air.[3] From more than a thousand years later in the same country, we can see another headrest, from the third or second century B.C.[4]

Lampshades Ivory lampshades have been made in two forms. One used the natural cylindrical (slightly conical) shape of the elephant tusk, especially from the large part at the base, where the tusk was embedded in the jaw. Application of abrasives then reduced the thickness so that the lampshade became more translucent. Moreover, the cylinder was pierced with many small holes, or cut through in imaginative designs, so that more light was emitted sideways and less heat would affect the ivory. The second form of ivory lampshade resembled a truncated pyramid with four joined flat sides. Here, too, the ivory was pierced or cut through.[5]

Panels The fruits of luxury in ancient empires introduced the use of ornamental ivory panel-sections, set into doors, room screens (partitions), thrones, and large furniture. Such decorative ivory panels are mentioned also in the treatment of the particular structures that utilized the panels, as in the next section below. Many ivory "plaques" that now appear as individual carvings—in collections, antique shops, and auction sales—are ivory panels that were long ago removed from old deteriorating settings, or were even picked off and stolen from settings that were still in serviceable condition.

Room Screens Chinese screen partitions and table screens sometimes featured a set of carved ivory panels, as well as artistic representations fashioned from gemstones and other precious materials.

Rope-Bedspring Tighteners Nineteenth-century American seamen used whalebone as the material in making a carved device for tightening the slack in the rope "bedsprings" still employed in that century. An example can be seen at the Mystic Seaport Museum in Mystic, Connecticut.[6]

Tables Tables adorned with ivory inlays date back to pre-Christian times. The Victoria and Albert Museum in London has a table and two chairs made of ivory to a substantial extent, with some parts made of wood covered with ivory veneer. Teapoys were small three-legged tables, popular in nineteenth-century India; the wooden structure was often decorated with inlays of ivory and mother-of-pearl (Fig.41). Although they often held a *tea* service, the actual origin of the term is believed to have come from the Sanskrit words for "three" *(tri)* and "foot" *(pada)*.

Thrones Probably the most lavish and impressive display of ivory furniture is seen in the thrones that once seated some of history's monarchs and church leaders at ceremonial functions. Early legends frequently mention "thrones of ivory," and there is a biblical account of a throne of ivory and gold for King Solomon of Israel in the tenth century B.C. (1 Kings 10:18; 2 Chronicles 9:17). From 701 B.C., the annals of Assyria's King Sennacherib tell that he received a tribute of ivory thrones and couches from Hezekiah, King of Judah.[7]

Some modern writers feel that the finest surviving ivory throne is the one at Ravenna in Italy—the magnificent throne of Maximian, Archbishop of Ravenna from 546 to 553. It was entirely covered with carved ivory panels, depicting biblical themes. Sadly, several of the ivory panels from the backrest are missing, probably stolen long ago; perhaps equally disturbing is the fact that they remain in collections where they were found in later times.[8] George F. Kunz judged that this throne was made in Egypt, although he mentions that O. M. Dalton, the British classical art critic, felt that it was made in Antioch (in Syria at that time).[9]

First used in 1671, the great coronation chair in Copenhagen for the kings of Denmark is made mainly of narwhal tusk.[10] In Rome there is the ivory Chair of St. Peter; and in Moscow's Kremlin there is the ivory throne of Ivan III.[11] In India, Rama Varma, the Maharajah of Travancore, had an ivory throne which he used at formal receptions (durbars). It was entirely covered with ivory that was decorated with dancing girls, pineapples, parrots, tigers, and dragons. At the beginning of the twentieth century, it was still in the old durbar hall in the fort at Trivandrum. His successor, Marthanda Varma, presented a splendid ivory throne with inset jewels and matching footstool to Queen Victoria, in honor of London's Great Exhibition of 1851. After being displayed at the exhibition, this gift was placed in the state apartments at Windsor Castle.[12]

Wall Plaques In appearance these were similar to ivory panels but were individual artistic pieces, not part of a larger object as in the instance of plain or carved panels. As in the case of artistic plaques for display on a table, desk, or shelf (see pp. 77–79), these wall hangings were usually framed. Although they served the same purpose as wall paintings, they were of course quite small.

Window-Shade Pulls Whaler seamen made these knobs from ivory or bone in the nineteenth century.[13] They were attached to the end of the window shade's pull string.

10 ❈ GAMES AND GAME ACCESSORIES

Because of the profuse variety of games that employed ivory, separate categories will be provided for puzzles and toys. As defined here, games will refer to those pastimes that establish a competitive engagement between opposing players operating under a prescribed set of rules, or at least those involving an attempt at attaining a good score if the game can be played by a single person.

Billiard Balls In Part One it was shown that billiard balls became one of the major uses of ivory. The importance of careful seasoning of fresh ivory, before cutting the billiard balls down to their final size, was briefly mentioned (see pp. 41–42). Although billiard balls made from mammoth-fossil ivory were sometimes produced in Russia, that very old ivory substance had lost much of its resilience and was therefore less commonly used than fresh ivory.[1] With ivory from recently deceased elephants, the straighter tusk of the female was preferred by some manufacturers because it facilitated the first machining operation of cutting out the spherical form. Also, some manufacturers claimed that the female tusks were harder, tougher, and more elastic.[2]

The hollow section of the tusk was sawed off, as well as the end point of the solid section. Then the remaining solid section was carefully measured, marked, and sawed into smaller sections, one for each billiard ball that was to be made. Each small section was mounted on a machine lathe, where a cutting tool gradually cut the spherical shape, usually while a jet of water sprayed on the ivory so as to prevent it from overheating and cracking. Then each sphere—slightly oversized so as to allow for subsequent shrinkage—was stored for seasoning, sometimes for several years. They were kept under conditions of constant temperature, because ivory expands and contracts during temperature changes. Stored ivory tusks have occasionally split apart with the explosive sound of a pistol shot when a storage warehouse's uncontrolled temperature suddenly increased; to a lesser extent, the same has happened while a billiard ball section of ivory was being stored or cut. Finally, the oversize ball is cut down to its required size. Holtzapffel recommended a multi-phase drying process: the roughly cut, oversize ivory ball is stored for 6 months, then cut down to $\frac{1}{16}$ inch (1.6 mm.) oversize. After another 6 months of shrinking in storage, the ball is cut down to $\frac{1}{32}$ inch (0.8 mm.) oversize. Half a year later, it is cut to its final size.[3]

While cutting a billiard ball, the machinist often uncovered defects caused by either imperfect tusk growth or some disease of the tusk tissue. Mere discoloration did not affect the playing quality of the ball, but it did lower the sale price. For example, about 1910 the finest quality ivory balls retailed at about $16 each, while highly discolored ones of otherwise equal quality sold at only $6 each. Any small soft areas that happened to show on the exterior surface would seriously diminish the ball's value; and any noticeable cavity, no matter how small it might be, rendered the ball totally unacceptable.

Expert billiards players demanded that the center of each ball be the original center of the tusk section that was used in making the ball, because that point would be the ball's center of gravity. Such a perfect sphere will run true when struck directly by the cue stick or by another ball. To test whether the finished ball met this expectation, some highly concerned manufacturers immersed each ball in a cup of liquid that was heavier than ivory (usually mercury, plus some benzol). If the floating ivory ball suddenly and vigorously gyrated before coming to rest in a "dead float," its center of gravity was not in the spatial center of the ball; that ball's value was diminished.

As a final note on the very critical standards imposed by expert players, we learn that at the great billiards tournaments of the past the room temperature was expected to remain at an optimum 72 degrees Fahrenheit (22° C), so that the ivory balls, subject to expansion and contraction from temperature conditions, would remain at a standard size. An old billiard ball is shown in Figure 45; its dried-out surface has numerous fine cracks from old age, but its condition has undoubtedly been aggravated by continual battering in play.

The standard size of billiard balls, however, varied from country to country. Also, the number of balls used changed in later times from four to three. This reduction served to ease the heavy drain on ivory tusks being exerted by the use of ivory for this game and for a vast quantity of other ivory products. But with the growing popularity of pocket billiards (pool), the requirement of sixteen ivory balls (even though they were of smaller size) led to recognition of the need for a substitute material. The synthetic balls arrived none too soon; by the early 1900s, the demand for billiard and pool balls had become excessive.

In 1920 in the United States alone, three million people were estimated to be playing either pool or regular billiards each day on 300,000 tables in 50,000 rooms. For newly added tables or for ball replacements, 60,000 new balls were needed each year. To supply the ivory, thousands of elephants would now have to be slain, in addition to the ivory taken from elephants dying naturally. The early chemical substitutes were satisfactory for pool, but not for billiards, the game of experts. The world's leading manufacturer of billiard balls offered $50,000 and royalties to the first inventor of a "perfect substitute" for ivory billiard balls, and even offered to assist the inventor in his research efforts.[4] Ultimately, various substitutes were accepted by manufacturers and the players.

Finally, it is of interest to note that when ivory billiard balls were replaced by synthetic ones, the ivory balls sometimes gained a second life—many were cut up and used for carving small articles, while some were painted with a map of the world (see Fig. 44).

Billiard Cues   Cues were sometimes decorated with ivory inlays or even one solid section of ivory. It is likely that some collection of early cues has one that is entirely of ivory, made of two or more joined lengths. A one-piece ivory cue may have been made from a narwhal's long straight tusk (as were spears).

Checkers (Draughts)   Playing pieces and the remnants of a checkerboard were found in Egyptian tombs of 1600 B.C. A hieroglyphic papyrus written a little later (now in the British Museum), preparing the Revenue Chancellor for his life in the afterworld, shows a drawing of Chancellor Ani playing at the game of checkers. About a thousand years later, early Greek writers mention the game. When were ivory checkers introduced? Barnett says some were found at the site of Megiddo in Palestine (1350–1150 B.C. period).[5] But Loud's earlier report on the ivories mentioned only "game markers," and no checkerboard.[6]

As compared with chess sets, relatively few checker sets have been made of ivory; complete sets seem to be rare. Ornamented checkers usually had geometrical designs or animal figures, painted or incised. Ivory checkers from the twelfth century A.D. can be seen at the British Museum.[7] The finest examples of pictorially incised ivory checkers are thought to be the ones from that century. Figure 45 includes an ivory piece that may have been a checker.

Chess   Miscellaneous ivory chess pieces from the eighth century have been found in India, a discovery not surprising when we consider that the earliest form of the game is thought to have originated in that country. European chess pieces of ivory can be dated back to the eleventh century. The British Museum has the most complete set of ancient ivory chess pieces, thought to have been made in the twelfth century. They were found in 1831 on the Isle of Lewis in the Hebrides, off the northwest coast of Scotland. The "Lewis chessmen" are of walrus ivory, numbering several dozen miscellaneous pieces, they but do not constitute one complete set. It is thought that they were carved in either Scotland or Iceland.[8]

In the sixteenth century, craftsmen in Italy made finely carved sets expressing great artistic sensitivity. During the following two centuries, Italy, France, and Germany produced magnificent ivory sets for wealthy patrons. These chess pieces were carved so lavishly and intricately that they were rarely if ever used for actual playing. French chess pieces—especially those made in the nineteenth century—sometimes portrayed well-known political and military personalities of the time. In England, large-scale production of ivory chess sets began in the early part of the nineteenth century; ivory chess pieces had first been crafted there in the twelfth or thirteenth century.

Although black and white were the standard colors for the opposing groups of chess pieces, most Victorian Englishmen preferred the red and white ones that came from India. Some English carvers introduced political and religious distinctions into the opposing groups of playing pieces. For example, one craftsman carved the white ones as Christian crusaders, and the red ones as "enemy" Saracen Moslems.[9]

Nineteenth-century India produced a vast number of ivory sets, varying greatly in the size of the pieces, as well as in the quality of carving. A Chinese specialty featured a circular base supporting a pierced ball, above which was a particular chess figure. Sometimes the pierced ball (for each of the thirty-two pieces) was the familiar Chinese puzzle ball, with one or two concentric balls inside.

In that same century, French carvers were producing highly regarded chess sets in Dieppe, including the unusual *pique sable* sets—literally "sand stick" sets. Those slender ivory chess pieces were atop long pointers which could be stuck into the sand at the beach. Players sat down on the sand and ruled off a crude chessboard of grooved lines for the squares. Among the more affluent players, when traveling by coach or rail, a thick square mat with ruled squares enabled them to play chess in transit; the ivory pointers were pushed down into the mat.[10] These sets of spiked chess pieces originated before the French Revolution.[11]

At one time the large upper teeth of the hippopotamus were frequently used for chess-piece carvings. Generally, however, carvers preferred making the entire set of playing pieces from a large single tusk of an elephant because of the difficulty in matching the surface appearance of pieces from different tusks.

In addition to the playing pieces, chessboards too were sometimes made of ivory. Although the inlaid squares were usually plain (of two contrasting colors), they were sometimes decorated. As an outstanding example of ivory art in a chessboard, the Bargello Museum in Florence, Italy, has a fifteenth-century French chessboard on which the sixteen dark squares are painted with detailed geometrical designs; the board's four side-margins are finely carved in relief with panels portraying historical themes.[12] India, already mentioned for its ivory chess pieces, made attractive wooden chessboards with ivory-inlaid squares. Some ivory chess pieces are shown in Figure 45.

Cribbage Boards Ivory cribbage boards of walrus tusk were a favorite work of Alaskan Eskimos, made especially for the tourist and export trade. Therle Hughes's book on small antiques shows a fine machine-turned cribbage board in circular form standing on six ivory legs; it was carved by Nathaniel Engleheart in the 1840s.[13] Early Eskimo cribbage boards of ivory have also been credited to a carver named Angokwazhuk, who became famous as "Happy Jack, the Ivory Carver" at Nome. He died in the influenza epidemic of 1918.[14] A Chinese cribbage board is shown in Figure 45.

Dice and Dice Cups The earliest dice (along with some kind of game board) were found in Egyptian predynastic tombs (before 3100 B.C.). In the eighth century B.C., Greece's epic poet Homer tells us that Penelope's unwelcome suitors passed some of their leisure time by "sporting with ivory cubes."[15] Martial, a Roman poet of the first century A.D., comments on the game played with "ivory knuckle bones": "When no one of the bones stands with the same face as another, you will say I have given you a great present [good luck]."[16] In one version of the game, either three or four dice were used; the winning throw required that different numbers come up on each of the thrown dice.

Used either alone or as accessories, dice have retained great popularity from ancient to modern times for providing the chance feature in games. Until replaced by synthetics, ivory was the preferred material, with bone serving as second choice. Ivory and bone dice and a pair of ivory dice-cups (for shaking the dice) are shown in Figure 33.

Dominoes Domino pieces were usually made of wood. Although the finer sets used ivory or bone as the top layer, glued to a wooden base (Fig. 33), some sets were undoubtedly made entirely of ivory or bone.

Eskimo Game In one of her books on Alaskan Eskimo art, Dorothy Ray tells of an Eskimo game called Tingmajuk that was played with flat pieces of ivory in the shape of birds.[17] Later, Carson Ritchie added information about this game, which he calls Ting Miu Jang. It uses fifteen to twenty-five small carved-ivory figures, which might be of a seal or other animal. Two players take turns throwing the pieces upward and letting them fall upon the table or floor. The thrower keeps the ones which point directly toward him. (There could be cause for argument here.) Each player throws only the pieces which remain. The winner is the player with the most pieces after a pre-agreed number of throws by each player. The British Museum has a set of these pieces.[18]

Game Boards (No Known Method of Play) The method or principle of play is not known precisely for some ancient games. The treatise of Howard Carter, one of the discoverers of the fourteenth-century B.C. tomb of the Egyptian King Tutankhamen, illustrates two small pocketsize gameboards of ivory 3 squares wide and 10 squares long. The playing pieces were inside the boxlike board. Each player had 5 pieces of his own color; the moves were apparently determined in some chance manner. A third board is shown, of medium size, but only the 30 squares are made of ivory. Another illustration of this game shows a large board with the 30 ivory squares; the other side provides a 20-squares game—3 by 4 squares, with an approach of 8 squares leading to the center row of 4 squares.[19]

A treatise of the British School of Archaeology in Egypt has illustrated and further described those two games. The 3-by-10-squares game (called Sent) is simply a rectangle divided into 30 squares. Each of two players had 6 identical playing pieces of a distinctive shape. (Note that Carter's treatise tells of only 5 markers for each player.) A pair of dice determined the number of moves required, but the aim and strategy are still unclear. This game is thought to have been popular from the third dynasty (2600s B.C.) to the twenty-sixth dynasty (600s B.C.). Those games of 30 squares which had been discovered went to museums in Cairo, London (British Museum and also the University College Museum), Manchester, Berlin, Leyden, and New York. (In most instances, only the city is named.)

For the 20-squares game, the method of play is unclear. Three special squares (the first, fifth, and ninth in the long middle line) are indicated by distinctive inscriptions or coloring: an apparent beginning or end point, an apparent capture point, and an apparent forfeit point. The games of 20 squares (with the reverse side showing the game of 30 squares) went to museums in Cairo, Leyden, and New York. The treatise also mentions a 58-holes game, with uncertain method of play, made in Egypt some time between the ninth and the twelfth dynasty (2100–1800 B.C.).[20] The game has most of the holes in the form of a large number 8 and is shown in a different treatise.[21] A recent publication shows an ebony and ivory board with 3 by 10 squares on one surface, called Senet in Egyptian, rather than Sent; the 20-squares game, on the reverse side, is called Tjau, translated as "Thieves."[22]

Golf Clubs Impressed by the resilience of ivory, some golf experts are said to have had an ivory disk inlaid in the head of their drivers. The author knows of no surviving examples of such collector's items.

Jackstraws Although the "pick-up-sticks" game almost always used strips made of straw or wood, nineteenth-century China furnished an interesting set of jackstraws in which each stick was made of ivory and was carved as a different farming implement or craftsman's tool.[23]

Mah Jong For this ancient Chinese game, the rectangular tiles have been made of various materials, ranging from plain to precious. During China's last dynasty (Ching, 1644–1911), a resurgence of interest led to production of sets made of ivory or bone, sometimes with a hardwood base (Fig. 33). Finally, the foreign craze for attractive and artistic Mah Jong sets resulted in using up nearly the entire Chinese supply of ivory, disorganized the ivory industry in China, and stopped the production of artistic carving in that country.[24]

Marbles This children's game dates back to very ancient times. Although most of the early marbles were probably made of clay—rather than of marble itself—it is of interest to note that ivory marbles are said to have been found in Egyptian tombs dated earlier than 3100 B.C. Presumably they were used for rolling or shooting in a marbles game. Three thousand years later, Roman children were playing this game. The present name of the game was assumed when attractive colorful marble became a choice material for the little spheres. Figure 45 shows an old ivory marble, now quartered for use in jewelry (e.g., for rings).

Markers Probably the earliest markers and counters were small stones, or perhaps beans and other seeds. As might be expected, precious materials such as ivory were employed for this purpose at some later date. Early ivory markers were found in Egyptian royal tombs of the period 2850 to 2600 B.C. Ivory continued to be a favored material for markers, almost always carved round and flat like coins. Ivory pegs became popular for game boards with holes.

Ivory markers in the Middle Ages were sometimes large and heavy; historical records tell that the losing player sometimes became so violently angry that he grabbed a heavy marker and bashed it against the head of his opponent! The eighteenth century initiated a great demand for game markers made of ivory (Fig. 33), as well as for ivory boxes in which they could be stored.

Playing Cards It may be surprising to learn that entire decks of thin ivory playing cards were made in India and Persia until the late nineteenth century.[25] Also to be noted are the small, bone playing cards made by French prisoners-of-war in England from 1795 to 1815.[26] It need hardly be said that a set of ivory playing cards would be a rarity now.

Poker Chips A set of game markers (see Markers, above) might be identical, or might differ in color or shape so as to identify the ones for opposing players, but poker chips are the same for all players, only differing (usually in three colors) to indicate the different betting values. Favored sets were made of ivory, until finally all sets were made of various composite substances. An ivory playing piece shown in Figure 45 was probably either a "red chip" for poker or a checker piece.

Poker Dice A variant of card poker used a set of five painted dice, which were sometimes made of ivory (Fig. 33).

Roulette-Wheel Balls It is generally thought that roulette originated in France in the seventeenth century. By the late eighteenth century, it had become a glamorous attraction in European gambling casinos (best exemplified in Monte Carlo). The game's only interest for us resides in the small white ball, which was originally made of ivory, but was later replaced by plastics. In America, where roulette has been most associated with Nevada casinos (especially in Las Vegas), collectors seem unable to obtain any of the ivory balls formerly used.

Scoring Devices The reason ivory was used for making scoring devices was probably its pleasant tactual feel and because it can be hand-carved with attractive designs. Some scorers are suitable primarily, or only, for one game, as in the instance of cribbage boards. Other devices, although usually intended for a single type of game, may serve for other games too. Figure 43 shows three examples: a pair of whist-game scorers from England (nine-point maximum score) made of ivory and mother-of-pearl, a 9999-point scorer from France of ivory and wood, and an eight-mark scorer of solid ivory with hinged tabs of ivory and wood for an Oriental game. It has been suggested that this latter scorer, with its five large marks and three small ones, might be for an old Chinese game called Li-Po (or Liu-Po).

Tops for Gambling Although gambling tops were probably introduced some time after toy tops (see Tops, p. 182) both kinds are very ancient; and both were sometimes made of choice materials such as ivory. The early Roman name "totum" is still used, meaning "all" (signifying "take all"). This top is also called tee-totum, because the letter "T" was inscribed on one of the four sides. The other three sides directed the player to either "Put" (again), "Take half" (or some specified number of coins), or "Nothing" (neither put nor take); only the letters "T," "P," "H," or "N" appeared on the sides. A Roman type of ivory totum is shown in Figure 33. (A similar top, made of vegetable ivory, was shown in Fig. 1.) Ivory gambling tops were later made with more than four sides to allow for more possibilities: P1, P2, T1, T2, and so on.

Gambling tops should not be confused with certain game tops that have only numbers inscribed on them and which are used in games where the player spins that top in order that it indicate how many spaces forward his own playing piece should be advanced on a game board. Those too were sometimes made of ivory.

11 ❊ HEALTH ARTICLES AND ACCESSORIES

Ivory has served health needs and practices more frequently than might have been supposed. More specifically, a few of those animal-tooth articles have serviced, and even replaced, our own teeth.

Artificial Teeth Because hippopotamus ivory did not yellow with age, it became a preferred kind of substance for making artificial teeth before synthetic substitutes replaced ivory for this purpose. Hippo-ivory teeth were made as far back as the Roman period.[1] In modern times, Kunz claimed that ivory was not particularly suitable for artificial teeth and had rarely been employed for that purpose.[2] Nevertheless, we learn that an eighteenth-century advertisement offered these advantages for the wearer of ivory teeth: "People of first fashion may eat, drink, or talk and show their teeth without hesitation."[3] In a personal letter to the present writer, Norbert Beihoff mentions that President George Washington was said to have had a set of ivory dentures, in addition to the wooden ones he was thought to possess. In 1980, however, the former Dean of the School of Dentistry at the University of California at Los Angeles, Dr. Reidar Sognnaes, publicly disputed that President Washington ever had false teeth made of wood. Dr. Sognnaes's examination of the dentures revealed that the artificial teeth were made of ivory and that subsequent discoloration had left them with the appearance of dark wood.

In London, ivory artificial teeth were furnished until 1875. A nineteenth-century reference tells that dentists were using walrus ivory—much more plentiful than the favored hippo ivory—for carving tooth replacements for their patients.[4]

Canes Consideration has already been given to canes as containers (see p. 98) and walking sticks as fashionable costume accessories (see p. 110). Especially when used as physical aids in walking, the terms "canes" and "walking sticks" are interchangeable.

The earliest use of ivory for canes seems to be uncertain. During the eighth to the tenth centuries, the Vikings carved entire narwhal tusks into ivory staffs.[5] It is probable, however, that some of the earlier ivory handles, now separated from their original article, once adorned a wooden cane. Nineteenth-century American seamen carved canes from whalebone and baleen.[6] The Los Angeles Museum of Natural History has a cane made of the vertebrae of a fish. Another is at the Mystic Seaport Museum in Mystic, Connecticut.

Doctor's Lady In China, the last dynasty (Ching, 1644–1911) featured that charming ivory female nude, lying on her side, so that the modest lady patient could merely point to the part of the model that corresponded to the location of her pain or other ailment. The earliest carvings of such medical models may date back to the 1500s or even earlier in China. Today, some think that the use was more of a philosophical attitude than a matter of prudery or modesty. Most of these ivory carvings lacked any substantial artistic merit.[7]

Infant Teethers A forerunner of the modern teething ring was the "gum stick" in the shape and size of a small slender finger, with which an infant could massage his or her gums. The material was ivory or other substances such as bone or coral. To prevent accidental swallowing of the teether, a larger object was attached at the other end; in the more expensive models, that object was a silver rattle or whistle. Later, the ivory teething-ring was introduced (Fig. 32).

Intra-uterine Device The endless surprises of ivory's abundant uses may again be illustrated, this time by a brief and incidental mention in *Time* magazine's Medicine section (May 26, 1980), which mentions that intra-uterine contraceptive devices (I.U.D.) were sometimes made of ivory.

Ivory as Medicine Just as surprising as the belief in the medicinal powers of ivory is the broad geographical extent to which this notion has gained acceptance. Ancient Chinese used a solution of ivory powder from narwhal tusk as a medicinal cure.[8] In northern India, a "tonic" medicine for strengthening the body was made from ivory dust.[9] That belief is not entirely unreasonable when we consider that ivory particles can serve as elements of food. As recently as 1920, it was reported that ivory shavings (left over from billiard ball cutting) were still being processed into an ivory jelly that was considered "highly nutritious for invalids."[10] In Japan, ivory powder from narwhal tusk was once the main ingredient in an oral medicine thought to be effective for reducing a fever. An old English treatise assures us that ivory powder in a cup of goat's blood will cure all pains, as well as expel "little stones" in the bladder.[11]

Along occult lines of belief, the National Museum of the Society of Antiquaries in Scotland has an ivory amulet $7\frac{1}{2}$ by 4 inches (19×10 cm.) which was once thought to be a cure for madness.[12] In Siam, ivory amulets were worn for protection against disease, and ivory dust was used as a curative agent.[13]

Further indication that the above examples are not isolated geographic instances is attested to by the fact that much of eighteenth-century Europe used ivory dust in solution as a remedy for fever, jaundice, liver and spleen diseases, abdominal disorders, and to stop hemorrhages.[14] This latter belief about bleeding is similar to the beliefs that walrus-ivory hilts on daggers could stanch the flow of blood and also induce the faster healing of the wound if the ivory was applied to the wound. Some thought that walrus ivory could ward off bodily cramps; others thought that it could prevent the deadly effects of poison. Boar tusk was once considered to be a remedy for lung and blood disorders.[15]

There are several examples of ivory being used to warn against a medical danger. Near Eastern Moslems once believed that if walrus ivory became moist when it touched food, the food was poisoned. Elsewhere, a similar belief held that if a narwhal-ivory wine cup began to "sweat," the wine within was poisoned. Malayans thought that hornbill ivory would change its color in the presence of poisonous foods or drinks; therefore this substance was favored for finger rings, probably because the ringed finger is close to food and drink.[16]

Massagers An eighteenth-century massaging device from Peking can be seen in the Chinese exhibit at the Field Museum of Natural History in Chicago. A wooden handle about 6 inches (15 cm.) long holds a grooved ivory massaging roller at one end; the roller is the size and shape of a lemon. Apparently, very few were made of ivory.

Medical Instrument Cases Lucian, a Greek satirist from Syria in the second century A.D., ridiculed certain physicians who concealed their ignorance of medical matters by making a great show of carrying an instrument box made of ivory: ". . . you are exactly like the quack doctors who provide themselves with . . . ivory cases for their instruments; they are quite incapable of using them when the time comes."[17] We can wonder whether any such ivory carrying-cases have survived.

Medicine Cases Ivory cases for medicines can be dated back to the first century A.D. The Roman poet Martial muses thus: "You see a medicine chest, the ivory equipment of a doctor's art."[18] More than a thousand years later, the Japanese inro would become famed as a small portable medicine case and would ultimately be carved from ivory and other attractive materials.

Pill Boxes The pill box was of course only a specialized form of medicine case, already considered above. Also, almost any small flat container could serve the purpose. In recent centuries both plain and decorative pill boxes were made in various countries; as with many other health accessories, the clean white appearance of ivory favored the use of that material. Figure 32 includes such an example, artistically painted in gilt and decorated with insects fashioned from colored glass insets.

Salve Containers Since ancient times special ivory containers have been used for salves and similar substances. Ointment containers from the period 1350 to 1150 B.C. were among the ivory artifacts excavated at ancient Megiddo in Palestine.[19]

Surgical Instruments Several kinds of earlier surgical instruments had ivory or bone handles. A scalpel with an ivory handle is shown in Figure 32.

Tongue Depressors A small ivory article in this writer's collection is thought to be a tongue depressor used by a doctor when examining the throat of an infant. This thin flat ivory piece is less than $2\frac{1}{2}$ inches (6.4 cm.) long (Fig. 32). Its actual purpose would seem uncertain.

Tongue scrapers have been made of baleen, the firm but flexible strips that hang down inside the baleen whale's mouth. Some scrapers may have been made of ivory.

Toothbrushes As in the instance of hat brushes (see p. 107) and hairbrushes (see p. 162), early toothbrushes had a handle or back that was sometimes made of ivory or bone. A 1913 report of the U.S. Department of

Commerce stated that China was making toothbrushes from walrus ivory.[20] Figure 32 includes an interesting American toothbrush. The back, securing the brush element, appears to be ivory or bone; more unusual is the plastic handle end, shaped to serve as a gum massager. Although this product is stamped "Patent pending" (and "Clean-Between Toothbrush"), this writer's search failed to discover whether or not a patent had been granted for this novel product.

Tooth-Cleaning Powder Boxes Kunz has stated that special ivory containers were made for tooth-cleaning powder.[21] Here, again, any reasonably sized ivory container would serve the purpose.

Toothpicks As would be expected, ivory toothpicks are among the very smallest of ivory products. Simple, solid examples in ivory were made in great numbers in the Orient. In some countries, ivory-handled toothpicks were crafted and a metallic pick extension was added. Figure 32 shows both types; the one with a gold pick is part of a protective ivory case. (See also Personal Groomer Sets for Men, p. 164.)

Toothpick Cases and Holders Occasionally, finger-shaped ivory cases were crafted for containing a small number of toothpicks. These slender cases were ideal as portable containers for purse or pocket.

By contrast, toothpick holders were made for the dining table. A small stand had one or two rows of holes at the top in which the individual toothpicks were inserted before a meal was served. They may have been an aspect of a gracious table setting in earlier times, but today no elegant table would provide toothpicks. Figure 32 includes an ivory holder for twelve toothpicks.

Vinaigrettes Despite its French name, this tiny container is thought to have been an English invention; they were very popular in England from 1780 to 1880. Vinaigrettes contained a small piece of sponge that had been soaked in vinegar; primarily they were carried in a lady's purse. Originally they were believed to be a preventive against disease, presumably guarded against by sniffing. But by the middle of the nineteenth century the function had been severely narrowed to providing smelling salts for reviving a fainted person, much as we now might apply spirit of ammonia. Although vinaigrettes were usually made of silver, a large number were crafted from ivory (Fig. 32).[22] By the end of the nineteenth century, frequent fainting diminished so much that production of those charming revivers seems to have stopped.

12 ❁ HOUSE STRUCTURAL ACCESSORIES

The use of ivory for structural accessories of a house could not begin to rival its use for constructing the house itself. A discussion of this latter, ancient claim will thus begin the following sections. Beyond this possibility, very few parts of a house have favored or even allowed the use of ivory.

Houses of Ivory The Bible tells that Ahab made a house of ivory (1 Kings 22:39), and also includes reference to "ivory palaces" (Psalm 45:8). If this is true, such buildings were probably wooden structures overlaid with small ivory panels throughout, possibly both inside and outside. In the early 1900s, Reisner, on the Harvard University Expedition, discovered what are considered to be the traces of Ahab's "ivory palace," deeply buried under Herod's temple. Then, in 1933, Crowfoot found there "a whole series of exquisite ivory plaques carved in low relief, some with colored inlays and gold leaf."[1] Those plaques may once have covered the walls of the house.

In some parts of Africa, before the value of ivory was realized, entire corrals used upright tusks posted in the ground. In his *Natural History,* Pliny (A.D. 23–79) mentions that cattle enclosures of ivory tusks were found in the domain of King Gulussa, regional head of a province in Numidia, North Africa, in the second century B.C.[2]

Doors Rarely, ivory plaques completely covered a wooden door. Frequently, however, small ivory inlays were used extensively against the background of richly designed wooden doors. The maharaja's palace in Mysore, India, had a pair of such doors (french-door style) beautifully inlaid with ivory designs.[3]

Door Knobs and Door Handles If ivory-covered house walls (see Houses of Ivory, above) actually existed, it is surely possible that those "ivory houses and palaces" had ivory door-knobs too. We do know that in recent history, Victorian England had door knobs made of ivory. Queen Victoria liked them so much she had them installed on the doors of the royal train.[4] Nineteenth-century whaler-seamen carved door knobs of ivory and whalebone. Those knobs may be the ones that were once seen on the doors of homes in Connecticut and elsewhere in New England; some were seen in the Johnson Whaling Collection at Princeton, New Jersey.[5] Curved sections of hippopotamus tusks were once used for carving fancy handles on doors. Now it is very difficult to collect an ivory door-knob or door-handle.

[136] Part Two : Uses of Ivory

Door Posts In the first century A.D., Pliny wrote that ivory tusks were being used in Ethiopia for the vertical side-supports of doorways.[6]

Electrical Insulators The Pratt-Read Company of Connecticut manufactured disc-shaped and cylindrical ivory insulators until items made of more satisfactory synthetics replaced them (Fig. 46). A special use of ivory electrical insulators can be noted in some of the early push-buttons used in wall switches. Ivory insulators are a good example of an ivory product that is now difficult to obtain but is nevertheless of very little value (because only a collector of utilitarian ivory products might be interested in them).

Faucet Handles Figure 46 includes a fine brass faucet with two ivory knobs constituting the major portion of the on–off handle. Its small size suggests that it may have been the faucet for a cask.

Servant's Call Signal This is a truly rare ivory item. At one time in England, ivory sometimes was used for the tube system that enabled the lady of the house to speak with a servant in another room. When she blew through her end of the tube channel, the air rushed along and out through the tube in the wall of the servant's room, causing two actions: a whistling sound and also (in case the servant was not there) a red peg to be projected outward about a half inch (1 cm.), acting as a visual signal that could be seen when the servant returned. Figure 46 shows the servant's end of the system, with the red "Please respond!" peg blown outward.

13 ⊛ MUSICAL INSTRUMENTS AND ACCESSORIES

For the ornamentation of musical instruments, ivory inlay was especially popular in India during the nineteenth century; it also has a long history elsewhere before that time. In addition to inlays, ivory veneer-coverings and even solid ivory have been used—the usage in each case depending on the particular musical instrument, and especially its sounding requirements. Fine examples are in London's Victoria and Albert Museum and at the Royal College of Music.

Castanets A pair of slightly concave shell-shaped clappers, operated by a single hand, usually relied on hardwood for producing their sharp clicking sounds, but many were made of ivory. The Louvre Museum in Paris has a pair of ivory castanets from Egypt dating back to about 2000 B.C.[1]

13 · *Musical Instruments* [137]

Pairs of elongated flat ivory clappers from ancient Egypt were shaped as human hands and had to be held in two hands, thus suggesting that they were used as clapping hands; they were no doubt louder than actual hands and demanded less effort. In London, the Horniman Museum has a pair from about 1450 B.C., and the British Museum has an earlier pair of clappers from about 2000 B.C.

Conductors' Batons Somewhere there may be a conductor's baton made of solid ivory. Figure 47 shows a pair with ivory ends purchased by this writer in a government curio shop in the Soviet Union; their origin is uncertain. If the simple one was intended for rehearsals only, the other one, with its musical symbolism carved in ivory, surely was worthy of being seen at the actual musical performance.

Drums Around the fifteenth century, small ivory drums began to appear in Ceylon.[2] They would of course have had to be made from the larger, hollow section of the elephant tusk.

Flutes Bone flutes are among the oldest of surviving musical instruments. Flutes made of ivory are mentioned by the Roman poet Virgil in the first century B.C. (in his *Georgics,* Book II). In China, ivory flutes were made during the Tang dynasty (A.D. 618–906). London's Victoria and Albert Museum has an eighteenth-century German flute made entirely of ivory. Usually only the mouthpiece of a wind instrument might be made of ivory. Figure 47 shows an old ivory mouthpiece for a piccolo made by the H. F. Meyer Company in Hanover, Germany.

Harps Part of an ivory lyre was found at ancient Megiddo in Palestine, from the period 1350–1150 B.C.[3] In Paris, the Louvre has a fourteenth- or fifteenth-century Franco-Flemish harp of solid ivory for the major sections. The ivory is richly carved in floral designs and New Testament scenes.[4]

Horns and Trumpets Although ivory horns have been used primarily for calls and signals rather than for music, they will receive some treatment here (see also Hunting Horns, p. 185). An early record of ivory hunting-horns can be noted in the biblical reference to them in Ezekiel 27:15, written in about the sixth century B.C. A trumpet, fashioned from ivory sections, made in Greece during the fifth or fourth century B.C., is now in Boston's Museum of Fine Arts.

Because the solid pointed end of the tusk has to serve as the beginning of the sound-transmission system, it must be scooped out. Hollowing out that portion was difficult for early carvers, lacking the electrically powered drills which now make the task so easy.

West African carvers made ivory horns blown from the pointed end and also long ivory trumpets that were blown from the side. In the Congo, natives made ivory trumpets almost six feet (1.8 m.) long. The Los Angeles County Museum of Natural History has an ivory trumpet from the Congo; it is 52 inches (1.3 m.) long.

Lutes and Other Stringed Instruments Lutes, guitars, and similar instruments were sometimes made partly of ivory, especially in the sixteenth and seventeenth centuries. Often ivory provided a complete veneer. Ivory and bone were sometimes used for the bridge of guitars, mandolins, and so on; also, the pegs and pins that attached the musical strings to the instrument were sometimes made of ivory or bone (Fig. 47). The same is true of the pin on the bottom of the larger stringed instruments that are held upright while resting on the floor, that is, the sturdy pin that keeps the body of the instrument above the floor.

There is an instance in which an American seaman removed the original tailpiece, bridge, pegs, and end-pin of his violin, and replaced them with ivory and whalebone parts.[5]

Organ Stops, Knobs, Tremolos In addition to the main use of ivory in organs for the organ keys' veneer-coverings, ivory was also used frequently for the handle parts, such as the stops, tremolos, and other regulator knobs on the instrument.

Piano and Organ Keys In the manufacture of the top coverings for the white keys on pianos, the first cutting provided elephant-tusk sections 4 inches (10 cm.) long. Some sections were then divided into pieces about 2 inches (5 cm.) long and $\frac{7}{8}$ inch (2.2 cm.) wide, to provide for the "heads" that would later receive the pianist's fingering. Other sections were reduced in width to provide for the "tails" that recede back from the heads. Each piece was then sawed into very thin slices (about 16 to 20 slices per inch). These thin heads and tails were soaked in water to remove extraneous animal matter, and then washed and dried out, with care taken to avoid possible cracking or warping. Then they were bleached in a hydrogen peroxide solution for a day or two, dried again, and then further bleached on glass panes under sunlight for a few days. The required number and kind of strips were glued in the correct positions to the top of the uncut wooden keyboard and then were polished. Finally, the keyboard was cut into the separate keys. Originally, the small front (vertical) surfaces of the keys were also made of ivory about 1 inch (2.5 cm.) high, but later they were made of cheaper materials such as Celluloid.[6]

It should be understood that some manufacturers probably used a simpler procedure than that outlined above. All, however, desired that the complete set of ivory strips for each keyboard should be of uniform grade; some preferred that the surface be without any visible grain. The manufacture of ivory keys for organs was basically similar to that described in the account for piano keys.

By 1900, piano-key veneers had become one of the leading uses of ivory. Ivoryton, Connecticut, and London were the main centers of production; Tonawanda, New York, and Cambridge, Massachusetts, became strong competitors to Ivoryton.[7] By mid-century, however, high-quality white plastics were rapidly replacing ivory on the pianos of most manufacturers. At the time of this writing, however, some of the finest pianos are still being made with ivory keys. An early piano key with ivory covering is shown in Figure 47.

Plectrums The ancient Greeks may have been the first to use ivory for the plectrums that picked the strings of the lyre.[8] In China, ivory plectrums were made during the Tang dynasty (A.D. 618–906). From the seventeenth century on, Japan used ivory for these picks.[9] Except for musical collections in museums, ivory plectrums have become so scarce that collectors seem unable to obtain them.

Rankets First mentioned in Austria during the sixteenth century, the ranket was a cylindrical wind instrument, short and thick, made of ivory or wood, with a long slender mouthpiece (as on the bassoon). It emitted a muffled and reedy sound and required a complicated method of fingering a group of holes around the outside of the cylinder.

Shaker Instruments Larger than, but similar to, an infant's rattle, the shaker instrument was sometimes made of ivory. The British Museum has one from India.[10]

14 ❂ PERSONAL COMFORT ARTICLES

Since ivory is such a personally endearing material, it is disappointing to discover that it has been adaptable to relatively few articles for personal comfort. (A much more prolific use was found in its application to personal grooming articles, as will be noted in that category; see p. 160.)

[140]   *Part Two : Uses of Ivory*

Back Scratchers, Head-Scratch Pins, Toe Scratchers   In ancient China, ivory head-scratch pins apparently predated the better-known Chinese back scratchers. When the back scratchers were crafted, they were frequently made of walrus ivory. Now, the scratch element—shaped like a hand with curved fingers—has become a recognizable item worldwide (Fig. 49). Two thousand years ago, the Roman poet Martial explained the value of "an ivory scratcher": "This hand will protect your shoulder blades when an irritating flea is biting you, or any insect fouler than a flea."[1]

Ivory back-scratchers became a Victorian period novelty, more ornamental than useful; the most favored ones had an ebony handle attached to the ivory scratcher.[2]

The most unusual kind of scratcher was the Chinese toe scratcher. One of these very rare articles came from Shanghai and can be seen in the Chinese exhibit at Chicago's Field Museum of Natural History. About 9 inches (23 cm.) long, it looks like a slender ivory letter-opener but is more sharply tapered at the end (the scratch point).

Fans   The first use of fans may have occurred when some primitive human used a fallen leaf as a cooling device; even now, the fan surface is referred to as the "leaf." Fabricated fans date back to about 2000 B.C. in Egypt, according to ancient records. The earliest surviving fan may be one from Egypt which dates from the fourteenth century B.C.; little more than the handle was found intact. The ancient Egyptian fan handles were of wood or ivory.

From China, the earliest surviving fan is dated from the second century B.C., but its sophisticated appearance suggests a long prior history of fan development there, thought to date back to about 1000 B.C. Near the end of the sixth century A.D., Chinese fans came to Japan and were further developed there.

In later times, fans in China and Japan became significant symbols in social and religious life. At the Chinese court there were detailed regulations concerning the types of fans appropriate for the different ranks of officialdom. Fans in Japan, besides being a cooling device, served also as a sunshade, as a device for beating time to music, as an object for symbolic gestures in the staged dance, as a facial cover of emotional expression, and, most elegantly, for using its spread surface to hold a personal object that was offered to someone, or for receiving such an object.[3]

Not only did fans sometimes serve to cover emotional expression, they became—especially in Europe—the only ivory product that has ever served to express a large variety of feelings: the angry flutter, modest flutter, confused flutter, merry flutter, amorous flutter, and many more. In Spain, the

33. *(Left)* Domino pieces; Mah Jong tile. *(Center column)* Dice in various sizes; hand-carved peg-board markers and leaf-design marker; machine-turned marker. *(Right)* Set of five poker dice; gambling top. *(Bottom)* Pair of dice cups.

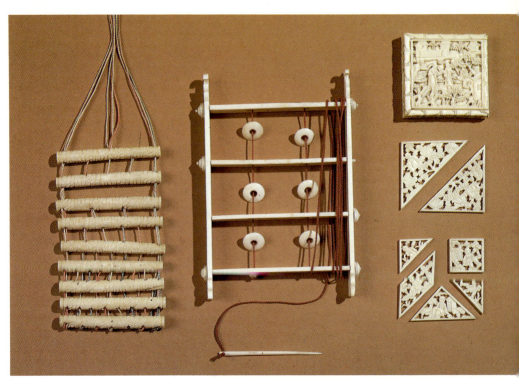

34. *(Left to right)* Chinese Rods-and-Cords puzzle; Chinese Ladder puzzle; tangram set.

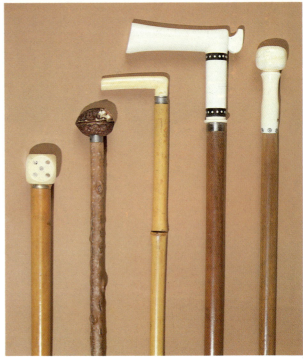

35. Ivory-handled walking sticks. *(Left to right)* Dice handle from France with gold inlay spots; English or American walnut handle; American sword cane (mid–1800s). Heavy-handled stick from Ethiopia; knobhandled cattle-prod stick, 67 inches, from Ethiopia (early 1900s).

36. (a) Busk. (b) Shoe buttonhook. (c) Closed-end buttonhook. (d) Glove buttonhook. (e) Clothes brush with ivory back.

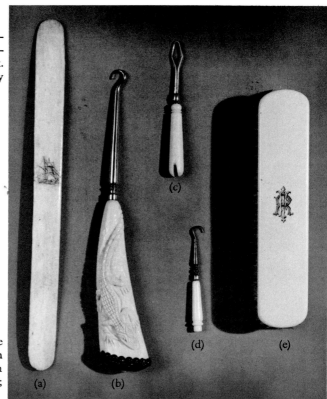

37. *(Left to right)* Glove stretcher; French hat brush with silver handle; Hat pin with ivory elephant figure; Japanese shoehorn.

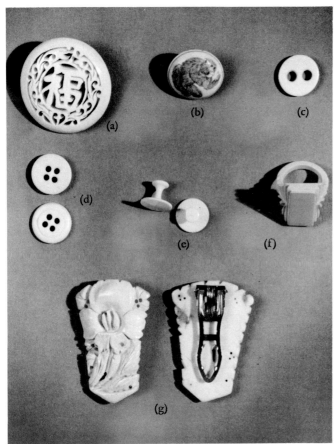

38. (a) Chinese or Japanese brooch. (b) Button with incised and painted monkey. (c) Plain two-hole button. (d) Four-hole buttons, showing front and back views. (e) Collar buttons. (f) Ring. (g) Pair of slipper clips, showing front and back views.

39. Models of the human eye (1700s).

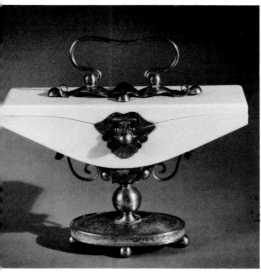

40. French jewelry case (*ca.* 1840).

41. Teapoy from India with inlays of ivory and mother-of-pearl, 20 inches high and 19 inches wide.

42. (*Top to bottom*) Marrow spoon; salad-mixing fork-and-spoon set; cake knife.

[145]

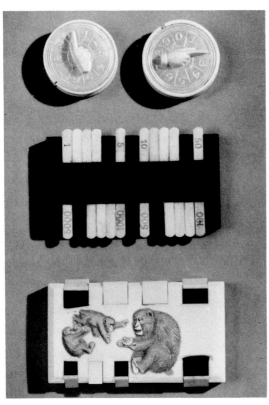

43. *(Top to bottom)* A pair of whist-game scorers; scorer with 9,999-point maximum; eight-point scorer.

44. Alphabet set from China; globe showing scrimshaw work on ivory ball.

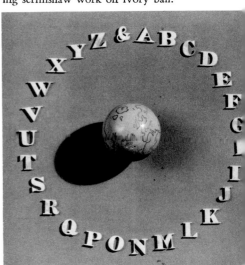

45. *(Top)* Old ivory billiard ball. *(Center row)* Two stained chess pieces from one set; white chess piece from different set; disc, probably a checkers piece or a poker-game chip; quartered marble. *(Bottom)* Cribbage board from China.

46. *(Top to bottom)* Four electrical insulators; faucet with ivory knob-handle; servant's end of tube communication system.

47. *(Top)* Two conductor's batons with ivory ends. *(Center, left to right)* Bridge peg and bridge for stringed instrument; ivory mouthpiece for piccolo. *(Bottom)* Piano key with ivory veneer.

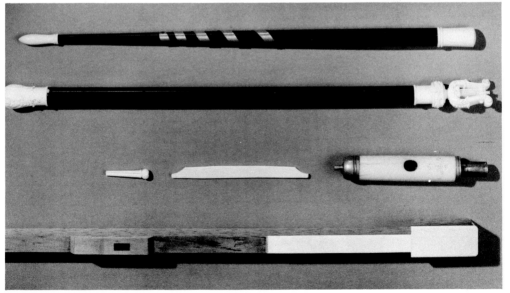

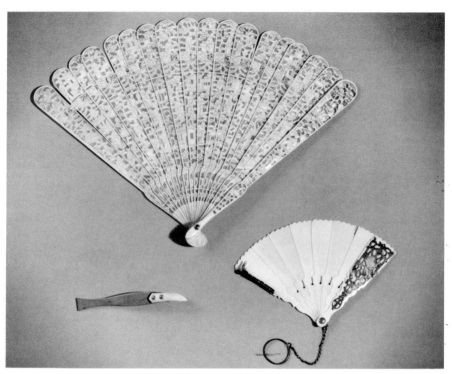

48. *(Top)* Chinese fan. *(Bottom)* Tweezers with ivory tips; dance fan from France.

49. *(Top to bottom)* Umbrella handle with four carved deer; parasol handle with "T" monogram, 11½ inches long; fly whisk with ivory handle; back scratcher with ivory hand.

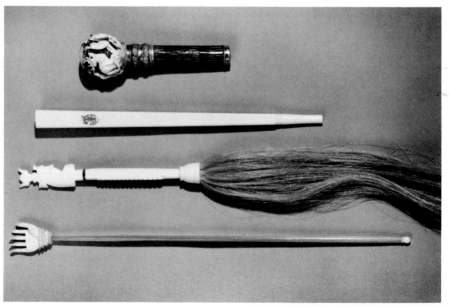

[148]

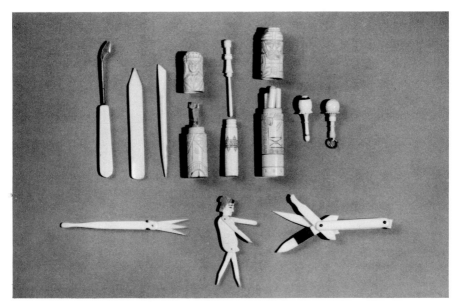

50. *(Top, left to right)* Three manicure instruments; perfume flask with inset glass bottle and stopper-applicator; ivory perfume flask and stopper-applicator; small manicure instruments in ivory case; two portable perfumers. *(Bottom, left to right)* Set of ear scoop and three toothpicks; Japanese manikin set; French three-piece set with an ear scoop (broken), toothpick, and combined fingernail cleaner and file (the opposite end holds an ivory cube with magnified see-through views of Niagara Falls).

51. *(Left to right)* Barber's neck-duster; fragment of an Alaskan Eskimo comb; hairbrush; Chinese hairpin; German straight razor.

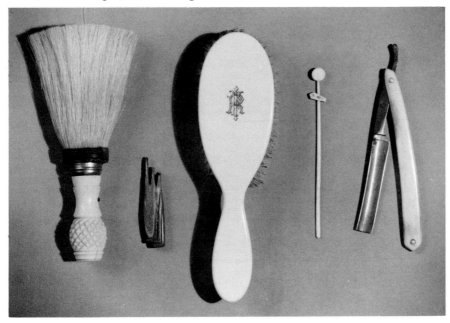

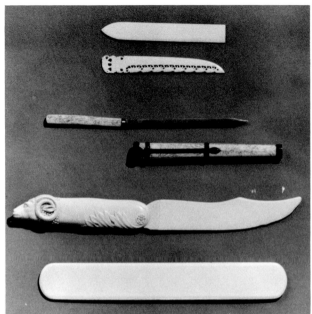

52. *(From top)* Four letter openers; paper creaser.

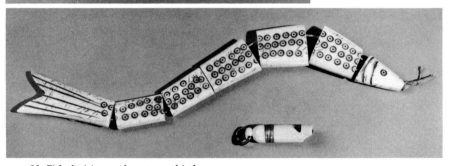

53. Fish decision-maker; toy whistle.

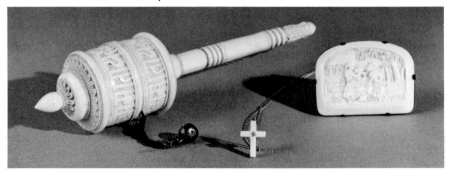

54. Buddhist prayer wheel; small ivory cross with gold nugget at center; European plaque with biblical scene.

55. Japanese ceremonial dagger and sheath; dagger with bird-head handle and sheath from Southeast Asia.

56. *(Left to right)* Pocket cigarette case; cigar box; cigarette case for a table or desk.

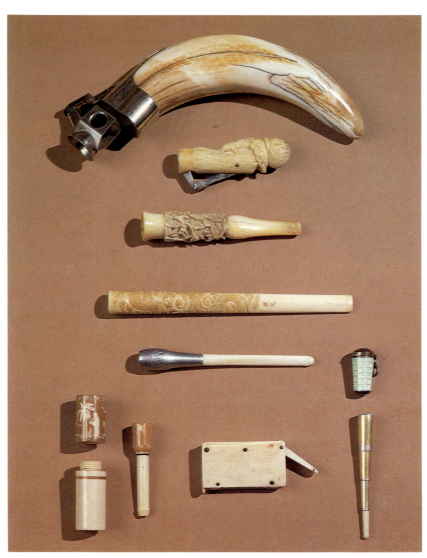

57. *(Top to bottom, left to right)* Two cigar-tip cutters; cigar holder; Chinese cigarette holder; cigarette holder with ivory mouthpiece; painted cigarette holder and case; match safe; collapsible cigarette holder with case.

(Facing page)
58. (a) Scratcher for attracting seals. (b) Small harpoon head, broken. (c) Fishhook, broken. (d) Fishing lure. (e) Armor platelet. (f) Arrow-shaft straightener.
59. *(Top to bottom, left to right)* Two paper rougheners; architect's ink pen; ink scraper; large penholder; pen-and-pencil tray with ivory inlay; in tray: ivory-handled pencil and small penholder; six-day diary; dance-card booklet with pencil and case; nameplates, old and new.

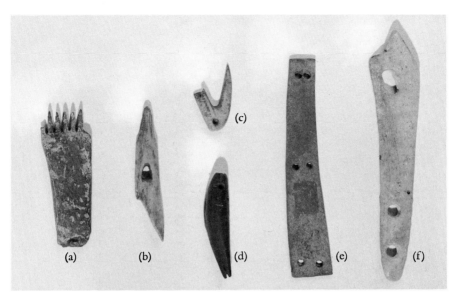

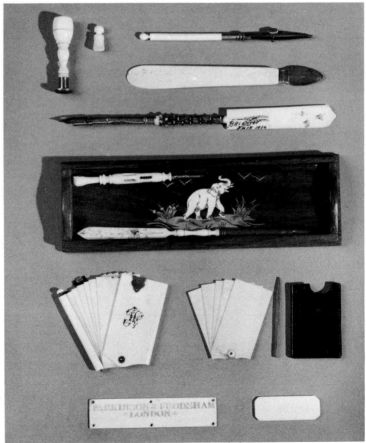

[153]

60(a). *(Top to bottom, left to right)* Ink-blotter holder; crocodile-design paperweight; seal-pigment box with dragon design; five seals, side view.

60(b). *(Left to right)* Top view of seal; underside view of four seals showing the stamp surface.

61. Inscribed gavel; bone section of a sled runner; celebrational baton.

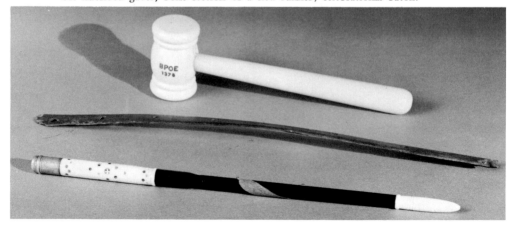

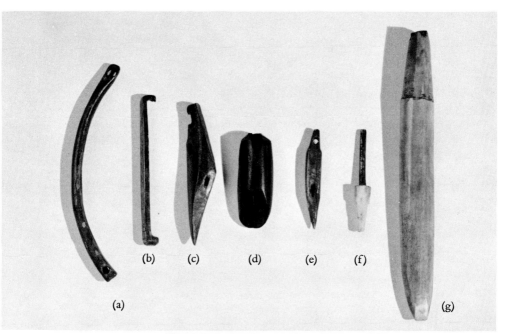

62. (a) Bucket handle. (b) Box handle. (c) Boat spur. (d) Adz head, stained black. (e) Awl. (f) Gouger chisel with ivory handle. (g) Ice pick.

63. *(Left to right)* Pick-and-whisk broom combination; pocketknife with handle carved as human figure; small corkscrew; folding ruler.

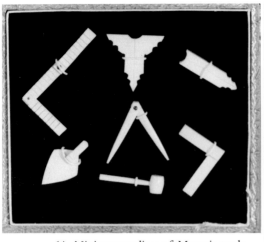

64. Miniature replicas of Masonic tools.

65. *(Top)* Lady's portable kit, including a refillable pencil, a bodkin, thimble, and needlecase. *(Bottom)* Latch peg for cardboard case; ivory coin from Cocos Islands.

66. Ivory products of uncertain use. (See p. 202.)

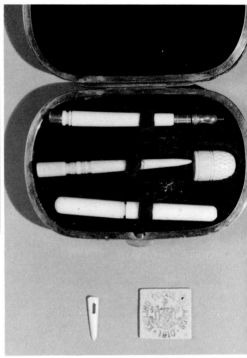

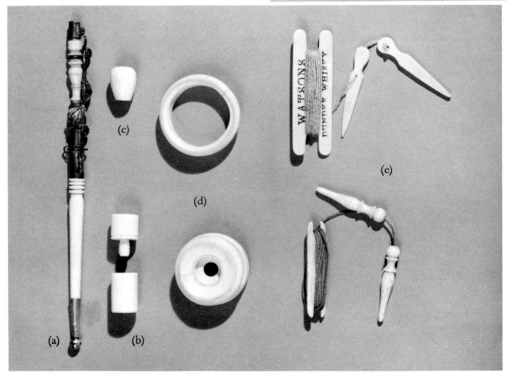

fan became the language of "silent conversation" among engaged couples during the seventeenth and eighteenth centuries, because the couple was not allowed to be alone. A secret booklet for lovers provided fifty messages, including the following:

"I love you" (Fan is placed near one's heart)
"I promise to marry you" (Open fan is slowly closed)
"When may I see you?" (Closed fan is placed against right eye)
"Do not betray our secret" (Open fan covers the left ear)[4]

Of course, ivory was not a unique material for any of the fans mentioned thus far. The early fabricated fans were of fixed form, now called rigid or screen fans: the handle supported a shaped paper or cloth form within a rim of wood or other substance. The handle, originally of wood, was later made of ivory in the more expensive fans. In one kind of ceremonial fan, the entire flat surface looked like a thin ivory sheet; actually, it was a plain weave of horizontal and vertical *threads* of ivory. Cutting a very large number of fine ivory threads off the length of a section of tusk was a painstaking and time-consuming task of craftsmanship. An exquisite ceremonial fan of this kind, made in the Chien Lung period in China (1736-1795), can be seen at Chicago's Field Museum of Natural History. (It was described in Part One, p. 17.) A similar fan, from India, made of woven ivory and silver threads, is at London's Victoria and Albert Museum.[5]

Next came the folding fans, in which a creased semicircular leaf (paper or firm cloth—plain, painted, or printed) was supported by a series of flat thin sticks that projected outward from a common anchoring point. The two end-sticks, or guards, were thicker, and the top one was later embellished elaborately. Some fans had sticks made of ivory, which were carved, painted, or pierced. Folding fans are thought to have been invented in Japan sometime between the end of the seventh century and the beginning of the tenth century. From Japan, they came to China. A Chinese record of A.D. 960 tells that when folding fans were first shown in China, they caused laughter and ridicule.[6] The earliest surviving example of a folding fan is said to be a paper-leaf fan dating from the twelfth century.

In the third major development of fans, the continuous leaf was eliminated by adding to the guards further sections of gradually increased width, so that they constituted an outspread arc of separate fan leaves. It is conjectured that this kind of fan, too, may have originated in Japan, around the twelfth century.[7] In the West, this kind of fan became known as a brisé fan, from the French term meaning "broken," because the fanning surface was now broken into separate sections; the fan sections, or leaves, were held together

[158] Part Two : Uses of Ivory

by a ribbon. Of special interest to us, this type of fan made it possible for ivory to be employed throughout the entire fan. Brisé fans achieved great popularity in Europe at the end of the seventeenth century, and then suddenly became highly fashionable a century later when they were shown and used by Madame Pompadour, the resident mistress and political confidante-advisor of Louis XV; she was fascinated by ivory, especially ivory fans.[8] During the eighteenth and nineteenth centuries, China made large numbers of brisé fans of ivory (and other materials) for the Western markets.

Before the French Revolution, Dieppe was a prominent center for the production of ivory fans. For several centuries, ivory fans were particular examples of very fine French carving and artistic embellishment. Elite Europeans, especially in the eighteenth and nineteenth centuries, seemed to regard fine fans not only as fashionable costume accessories, but even as socially required in certain situations. In England, although ivory fans were preferred over the ones of cheaper materials, only the wealthier Victorian ladies could afford them.

The production of beautiful fans declined steadily after the nineteenth century as women's roles approached those of men and as costume conventions eased. According to G. W. Rhead, "The last stage of the fan during this foolish, frivolous, fascinating eighteenth century [in France] was that of a gradual dwindling into nothingness."[9]

Figure 48 shows two ivory fans. The larger one is Chinese, with seventeen ivory leaves plus the two guards, each one with different scenes; all, except the two guards, have cut-through filigree ivory threadwork. The smaller one is a French dance fan (probably 1810 to 1840), with silver facing on the top guard, and a lock clasp with a tiny pencil inside it; the connecting ribbon is missing. Carried on mademoiselle's finger by the chained ring, she could write the names of the fortunate young men to whom she had promised a dance; the pencil writing on the ivory leaves could easily be erased.

Spectacular pictorial displays of beautiful ivory and other fans of the past can be seen in two works previously cited, one by G. W. Rhead and the other by Nancy Armstrong.[10]

Fly Whisks Valued ivory whisks were usually reserved for preventing flies from desecrating a religious or other revered object, rather than for warding off annoying flies; they are therefore treated in detail under Religious Articles (see Flabella, p. 171). It should be noted, however, that secular fly whisks with ivory handles have a long history (Fig. 49). Some were found in the excavations at Nimrud in present Iraq, from the Assyrian period 1200 to 600 B.C.

Parasols and Umbrellas Considered literally, our terms for these articles—parasol (French: *para,* against; *sol,* sun), and umbrella (Italian: *ombrella,* little shade)—reflect their ancient use for protection against the sun rather than rain. In perhaps the earliest use, in Egypt, the need for protection against the blazing sun is understandable. Not until recent centuries has our term "umbrella" come to signify a rain-oriented usage. Therefore, as concerns our interest in ivory's role, a distinction will be made between parasols as sunshades and umbrellas as rain guards. The parasol was destined to become an article of ceremony, fashion, elegance, and charm, but the plain and lowly umbrella became one of the most provocative and controversial articles of any that were sometimes made, at least partly, of ivory.

When were ivory handles first used? It seems uncertain as to whether the separated ivory handles discovered among ancient remains—in Egypt, Phoenician regions, Assyria, China, and India—were once attached to sunshades. The earliest written record about an ivory-handled parasol may be the one by the Greek poet Anacreon (500 B.C.), in which he scornfully refers to a certain newly rich fellow: ". . . now he goes in a coach, wearing earrings of gold . . . and carries an ivory sunshade as though he were a woman."[11] By 1700, umbrellas had begun to appear in England; soon thereafter they came to America. The ribs were commonly made of whalebone, later to be replaced by steel ribbing; ivory handles were infrequently seen. For the first three-quarters of that century, the umbrella was regarded as a curiosity. Even as late as 1800, Britons risked public scorn when carrying an opened umbrella.[12]

The use of ivory for umbrella handles might have been limited by the fact that the umbrella need be only a plain article that would be useful for a few minutes, or perhaps an hour, of bad weather, during which time hardly anyone could be expected to be looking at people's umbrellas. On the other hand, the parasol flourished in good sunny climates, where it was certain to be useful and seen for an entire afternoon or even longer. As a fashionable costume accessory for women, it favored an attractive handle of fine materials such as ivory, sometimes embellished with inlays of silver, gold, or precious gemstones. Even more important to the ladies, the sunshade offered an endlessly changing variety of coverings and countless designs in paper, cotton, satin, silk, and other materials. Ultimately, when the tattered cover and old parasol were no longer thought to be worth recovering, any handle made of ivory was likely to be removed, and then might someday make its way to a curio or antique shop. A modern collector browsing through a number of such shops should not find it too difficult to obtain an ivory handle that once belonged to a parasol. He should also be able to find an ivory umbrella-handle, though these are less common.

[160] *Part Two : Uses of Ivory*

Umbrella handles are usually larger (Fig. 49). A small number of prized parasols were made with a structural framework that was almost entirely of ivory or bone.

As had happened with walking sticks, one form of parasols and umbrellas underwent a period of gimmickry. Unique models (for people who could afford them) had a hollow handle for containing an additional article such as a perfume flask, writing utensils, stiletto, or telescope. (One of Queen Victoria's parasols may have been the only one ever made with a life-protection covering. Because someone tried to shoot her down, she had a parasol lined with bulletproof chain mail. It is now in the London Museum.)

Ivory handles for parasols and umbrellas were fated to become a thing of the past, partly because of changing fashions and partly because of the rapid increase in the cost of ivory.

Ivory-handled parasols and umbrellas can be seen in many museums throughout the world. In England, large collections can be seen at the London Museum, the Victoria and Albert Museum in London, the Birmingham City Museum, and the Gallery of English Costume at Platt Hall in Manchester. In the United States there are collections at the Museum of the City of New York and at the Essex Institute in Salem, Massachusetts.

Tweezers At least a small number of tweezers were crafted with ivory tips. Figure 48 includes an example; such an item could of course serve for removing a sliver or ingrown hair, and for plucking eyebrows. Small ivory-tipped forceps were also used in chemistry laboratories for picking up small metal pieces.

15 ⊛ PERSONAL GROOMING ARTICLES

A fairly large number of direct or indirect usages of ivory suggested themselves for articles related to personal grooming. Indirect usages are best exemplified by containers, a form of utility that is seldom closely related to the particular contents. Other indirect usages in grooming can be noted in the ivory handles of some of the implements.

Barber's Neck-Dusters One of several ivory-handled implements that were once seen in barbershops (Fig. 51), its application came at the end of the haircutting operation, when the barber dusted off the small cuttings that adhered to the neck, and also when he used it for applying talcum powder. By the end of the nineteenth century, the use of ivory for the handle was becoming less frequent and was soon entirely replaced by plastics.

Blemish Removers It has been reported that the Romans used ivory powder mixed with "Attic honey" for removing facial blemishes.[1] Perhaps it was used as a blemish covering, similar to the cover creams that are in use today.

Cold Cream Boxes Ivory containers are said to have been used for cosmetics such as cold cream.[2] As was mentioned earlier, the present writer questions the suitability of any ivory container for liquids, or even semi-liquid substances, because they slowly begin to dissolve away the inner surface of the container.

Combs Ivory combs were found in Egyptian predynastic tombs (before 3100 B.C.); animal figures were portrayed on some of the ivory handles.[3] From the period 1350 to 1150 B.C. in the ancient town of Megiddo in Palestine, the archaeological findings included ivory combs; some had a row of teeth on both edges. Ivory combs from China date to before 1000 B.C. The Louvre Museum in Paris has two ivory combs, believed to be either Assyrian or Phoenician, made sometime near the eighth century B.C.[4]

Along the Guadalquiver River in southern Spain, the remains of ancient cremation sites have yielded ivory combs, among other artifacts; the dating is thought to be as early as 700 B.C. The style appears to be Phoenician, possibly brought over from the Carthage-Libya area in North Africa. Because the ivory combs were the most frequent artifacts found, they are thought to have had some special burial significance. Various animals are portrayed; in one instance, a lion is attacking a gazelle.[5]

From a time almost a thousand years later, ivory and bone combs were found in tombs of Roman England; these, and subsequently combs from the Anglo-Saxon period, were sometimes richly carved with human figures, interlacing scroll ornamentation, and other designs. One of the finest examples of surviving decorative combs is a ninth-century German carving now at the Schnütgen Museum in Cologne.[6] Chinese women often adorned their heads with a number of ivory combs worn together. By the thirteenth century, England was producing ivory combs in large numbers. Later, the English firm of Puddlefoot-Bowers-Simonett manufactured ivory combs for a period that stretched across 250 years.[7] London's Victoria and Albert Museum has a fourteenth-century French comb showing lovers holding garlands in a garden; both of the long edges have a row of ivory teeth, one set being so fine and slender that it has about four times as many teeth as the opposite set.[8] The same museum has a delightfully picturesque comb from the fifteenth century showing a "Fountain of Youth" scene.[9]

India furnished a variety of ivory combs, including pocket combs that folded up inside a handle. Near Bombay, the city of Poona had a separate factory manufacturing ivory combs. But by the 1880s India's ivory combs were gradually being replaced by the new ones made of Celluloid.[10] A very unusual form of comb was the Chinese circular comb, cut from a thin cylindrical section about 1½ inches (3.8 cm.) high around a small tusk. Its use seems odd: "for scratching the head during a shampooing" according to the caption where it is exhibited at Chicago's Field Museum of Natural History. That exhibit also shows folding combs and moustache combs.

Another specialized kind of comb was used on children: the "nit comb" had teeth and spacings so narrow that tiny lice and their egg deposits could be combed out of a child's hair and scalp. In the United States, most of these ivory combs were made by the Pratt-Read Company in Connecticut. The first American combs of ivory were made by Andrew Lord in 1789 at Centrebrook in Connecticut. He used a handsaw to cut the flat plates and the teeth. Later, with fine machine-milling, some ivory combs had as many as 49 teeth per inch (2.5 cm.). Depending mainly on a comb's length, plain ivory combs sold for 25 cents to $2.50 in the early 1900s.[11]

Figure 51 includes a remaining fragment of a narrow type of Alaskan Eskimo comb probably made from walrus ivory; some simple decorative incising can still be seen.

Eye-Paint Containers Small containers for ladies' eye paint were made of ivory in India during the nineteenth century.[12]

Face-Powder Boxes A face-powder box made of ivory can be seen in the Chinese exhibit at Chicago's Field Museum of Natural History.

Face-Powder Brushes Ladies' soft brushes—some with ivory handles—for applying face powder (or probably for powdering any part of the body) were available some years ago. These were similar to the barbers' neck-dusters considered on page 160.

Hairbrushes Although it is likely that ivory-handled and ivory-backed hairbrushes were made in earlier centuries, the present writer has seen no mention of them before the nineteenth century. The same is true of the face-powder brushes discussed above, as well as of the clothes brushes described in the category on costume articles (see p. 106). Figure 51 includes a hairbrush with an ivory handle and back; this monogrammed brush is a companion piece to a clothes brush shown under the costume articles section (see Fig. 36e).

Hair Smoothers In some parts of India, thin slabs of ivory, about the size and shape of a paper creaser (see p. 190) were used by men for smoothing down their hair under their turban, so that the turban would not have to be removed.[13]

Hairpins and Hairpin Boxes Ivory hairpins were found in Egyptian predynastic tombs (before 3100 B.C.). China's first dynasty (Shang, 1500–1027 B.C.) produced ivory hairpins that were often elaborately carved, incised, and decorated.[14] About two millenia later, the wives of Viking seamen (eighth to twelfth centuries) were wearing hairpins carved from narwhal tusk. Figure 51 includes a Chinese hairpin of bone or ivory with an inscribed embellishment piece.

Special ivory containers for hairpins have been provided, but there could hardly be anything distinctive about a container for such a use.

Ivory Black as Hair Dye Ivory black was made by intense heating of ivory dust. One use of it was as a so-called hair restorer, but it was probably, in effect, a simple hair dye.

Manicure Instruments Various countries have furnished ivory manicure instruments, made either entirely of ivory or with only an ivory handle. Figure 50 shows both kinds. Some sets, usually sold in shops for men, included an earwax scoop (see Personal Groomer Sets for Men, p. 164).

Mirrors and Mirror Cases Until about A.D. 1500, most mirrors were a highly polished metal-alloy surface. Ivory handles for metallic mirrors were found in early Egyptian tombs and may date back to before 3100 B.C. One was also found long afterward on the island of Crete and is thought to have been made there around 1800 B.C.[15] Almost two thousand years later, in the first century A.D., India was producing such carved ivory handles for mirrors.

In medieval times, small round hand-held mirrors were sometimes encased in an ivory backing that was a fine work of sculptural art. The carved themes usually showed scenes from domestic life or from popular romances or poems. One of the favorites showed the charming "Siege Upon the Castle of Love," in which the knights outside were preparing to "force" their attention upon the highly interested maidens inside; the portrayal revealed little or no resistance. Another popular theme makes it evident that gentlemen and ladies played checkers and chess as opponents, more frequently than happens at present. Besides the small round mirrors, some ivory-backed mirrors were as large as 10 by 6 inches (25 × 15 cm.).[16]

Related to the hand-held mirror was the mirror case, a small flat container (usually for cosmetics) with an added feature of a mirror on either the top or underneath side of the cover. During the twelfth century, England was producing a substantial number of ivory mirror-cases; the same was soon true in much of Western Europe. It is often said that the fourteenth century produced the finest ivory artwork in mirror cases. Very fine mirror cases can be seen at the British Museum, the Victoria and Albert Museum, the Louvre, and the National Museum in Florence.[17]

Perfume Flasks and Perfumers Especially in the eighteenth century in France and China, large numbers of finely carved perfume flasks were made of ivory. The standard kind had a flat bottom and could stand up on a dressing table similar to modern perfume bottles. Simpler but smaller ones had a screw cap and could be carried safely inside a purse.

Smaller still were the ivory perfumers, one end of which was perforated and contained a small piece of cotton saturated with a few drops of perfume; the fragrance then continued to exude through the tiny holes.

Northwest of India, a similar but more dainty kind of perfumer was made by Kashmiri carvers—a flat ivory flower which had a single tiny hole in the center above the hollow interior. A drop of perfume was deposited into the cavity; when the ivory flower was worn on the wrist attached to a bracelet, the wrist surface warmth caused the perfume to gradually evaporate through the tiny hole.[18]

As a striking example of how thin a skilled artisan could cut (or grind down) a sheet of ivory, some perfume atomizers made in Ceylon were so flexible that they could be squeezed so as to expel a perfumed spray.[19]

Figure 50 includes two perfume flasks, one with an inset glass bottle and stopper-applicator; also a perforated perfumer with a metal connecting link, worn as a costume accessory; and another perforated perfumer, but with no connecting link, possibly intended for a purse (or perhaps a link was to be attached later).

Personal Groomer Sets for Men India, Japan, and several other countries made short slender earwax cleaners, or ear spoons, that had a tiny scoop at one end. Usually these were available in shops for men. They were often part of a two-, three-, or four-piece set, the pieces being hinged together at one end. Japan furnished a small, flat ivory manikin that had four folding limbs: two serving as toothpicks, a third one for cleaning the fingernails, and the fourth one for scooping out earwax. This manikin, a female figure, is shown in Figure 50. Also shown is a set with an ear scoop and three hinged toothpicks, and a French three-piece set.

Razors Men's straight razors were sometimes made with handles of ivory or bone (Fig. 51). That kind of shaving utensil was almost entirely replaced (except in barber shops) by the all-metal safety razor.

Shaving Brushes Fine shaving (lathering) brushes were sometimes made with ivory handles. Shaving brushes outlasted the straight razor, but their use declined greatly—along with the safety razor—with the introduction of electric shavers.

16 ❂ PUZZLES

Puzzles made of ivory, as distinct from ivory toys and games, are presented in this separate category because of their special problem-solving requirement. Toys lacking this feature are discussed under their own category (see p. 181). Games which, unlike toys and puzzles, involve competition between opponents, are also treated under a separate grouping (see p. 122).

Excluding the pencil-and-paper puzzles, most mechanical puzzles have been made of cardboard, wood, metal, or plastics; but many kinds were at one time made of ivory, although not in large quantities. China furnished entire cases containing only ivory puzzles. An attractively lacquered wooden box, about 12 inches (30 cm.) square and perhaps 6 inches (15 cm.) high, had an upper and lower tier with two or three dozen different puzzles made of ivory. Today, such a case, with its complete set of ivory puzzles in fitted compartments, would be a rare and prized kind of collector's item.[1]

The simplest kind of ivory puzzle consisted of a set of ivory markers or pegs that had to be moved around on a board in order to accomplish a specified task. In Peg Solitaire, for example, any peg had to be jumped over an adjacent one (which was then removed) until only one peg remained on the board, sometimes in the required center hole.

Many ivory puzzles required disassembly of the parts, or removal of one or more specified parts. Figure 34 shows, for example, the Chinese Ladder puzzle. By manipulation of the ivory needle and its attached string, we must remove the six ivory washers without unthreading them from the string, and keeping the end of the string attached to the bottom of the ivory ladder.

Related to the disassembly and removal puzzles were the various puzzle boxes, some made of ivory, that were difficult to open unless you could detect (or chance upon) the secret principle. Those ingenious boxes mainly required that some of the six sides be slid sideways, upward, or downward, in the exact predetermined sequence. Collectors now find it almost impossible to obtain an ivory puzzle-box.

[166] Part Two : Uses of Ivory

Figure 34 includes one form of the Chinese Rods and Cords puzzle (also known as Japanese Rods), which is quite different from what we might expect a puzzle to be. Rather than requiring manipulation toward solution, it merely invites understanding the puzzling construction by which the nine silk cords are threaded continuously through the nine ivory rods in such an ingenious way as to connect each two adjacent rods by a different number of cords (varying from two to nine cords). Three cords are yellow, three are red, and three are green.

Tangrams Perhaps these are the best known ivory puzzles; thus a special section is devoted to them here. The name is no doubt derived from "Tang," a Chinese dynasty (A.D. 618–906), and "gram," implying "picture." Whereas most puzzles can present only one problem, a tangram is a set of pieces that can be used for an almost unlimited number of problems. Similar to jigsaw puzzles, the flat pieces must be arranged to form a designated figure or picture. There are seven pieces, made by cutting a square into a set of geometrical forms: five triangles, one square, and one rhomboid. All seven pieces must be used for each tangram figure or picture attempted. The ordinary problem requires arranging the seven pieces to form the tangram (for example, an animal or familiar object) that is shown with the set.

Figure 34 shows an ivory container case with its seven pieces arranged in two squares. No two of the pieces have exactly the same artistic portrayal. The same is true for the six sides of the little ivory case.

17 ❈ RELIGIOUS ARTICLES

Religious art in ivory was discussed earlier, under the category of Artistic Ivories (see pp. 77–79). For a populace that was mostly illiterate, certain kinds of ivory art showing biblical events served as descriptive storytelling devices. Some of the Christian representations were effective in arousing deep compassion, more usually noted in the pictorial arts. Even the large scope afforded by a life-size marble pietà hardly exceeds the depth of grief evident in the small figure of Christ's mother in some ivory carvings.[1] Figure 54 includes a plaque depicting Abraham's servant receiving drink from Rebecca, with the gift of a bracelet in his hand behind his back (Genesis 24:22).

With an intent that was much less personal and subjective, special utilitarian ivory products have served a variety of ecclessiastical needs and functions. Most of these were in the service of the Christian church. Many had been secular types of products before being adapted to church use; a few kinds may have been uniquely ecclesiastical at first and then became part of the secular realm of ivory products. As will be noted below, most of the ivory articles in religious use were finely carved and embellished, reflecting a tone that was commensurate with the authority and dignity of the church.

Abstinence Plaques A few centuries ago the Chinese sometimes wore an ivory abstinence-plaque pendant to indicate to friends and associates that the wearer was making a religious fast and should not therefore be offered certain forbidden foods.[2] Most surviving examples of such plaques appear to be made of walrus ivory, although there are a few in existence made of hornbill.

Book Covers Ivory covers for some ecclesiastical treatises are treated separately (see Book Covers, p. 90). Examples can be seen in the Vatican Museum, in London's Victoria and Albert Museum, and in some other museums and several monastery libraries in Europe.[3] One reason given for the small number of surviving examples is that many of these book covers were later removed and were adapted to uses of a more visible nature, such as framed plaques for desks, shelves, and walls.

Crosses and Crucifixes Plain ivory crosses have been carved since early Christian times. Less frequently made was the ivory crucifix, showing the Corpus Christi (Body of Christ) on the cross. Sometimes the ivory figure was on an ivory cross, but usually it was on a cross of wood, thus reproducing the original cross of Christian history.

Rarely seen are ivory crucifixes made before the end of the Middle Ages, a period of more than a thousand years. The reason so few have survived is not well understood, considering the fact that a substantial number are thought to have been made before A.D. 1500. The scarcity of examples from England may be due mainly to the Protestant Reformation's anti-Pope revulsion that led to massive destruction of the accumulated Roman Catholic symbolic articles, especially during the mid-1500s. For example, the fourteenth-century "Christ on the Cross" in London's Victoria and Albert Museum may be the only surviving ivory crucifix of English origin from the twelfth century through to the fifteenth.[4] More puzzling is the scarcity of ivory crucifixes from those European countries that were free of the ravages and destruction of "popish articles"; ivory crucifixes had been carved profusely in France, Germany, and the Netherlands. Most ivory crucifixes seen today are from the seventeenth and eighteenth centuries.

A few art historians have thought that an ivory crucifix (now in Munich) was carved by Michelangelo, and that another one (whereabouts unknown) was carved by Cellini, but neither attribution has ever been substantiated. One of the greatest carvers of ivory crucifixes was Faistenberger (1646–1735). In 1681 this Austrian artisan, working in Germany, carved a crucified Christ that was almost 30 inches (75 cm.) long. It came into the possession of the Archbishop of Tours in France. Joseph Villerme (various spellings) was born in France in 1660 and died in Rome around 1720; he devoted his entire life to carving crucifixes of wood and ivory. Sadly and ironically, that devotion also led to his death, when he contracted a disease from a cadaver he was using as a model for a crucifix.[5]

In the seventeenth century, an ivory crucifix was carved in Ceylon and was sent by the Bishop of Goa (west coast of India) as a gift to the King of Spain. The figure of Christ was 3 feet (90 cm.) long and was carved so realistically that the depicted suffering on the cross caused some viewers to cry.[6] One of the most painfully realistic ivory crucifixes, showing the wounds bloodied red, is in London's St. James Church at Spanish Place; it is thought to be of Spanish origin from the seventeenth century.[7]

Pectoral crosses were worn by church officials, usually of higher ranks. These crosses were large enough to be seen easily from a distance. London's Victoria and Albert Museum has one of walrus ivory, an Anglo-Saxon carving from about A.D. 1100; it is $4\frac{3}{4}$ inches (12 cm.) long and 2 inches (5 cm.) wide.[8]

Finely carved ivory crosses and crucifixes can be seen in many museums (and some churches) in Europe, especially in England, Spain, Germany, France, and Italy.[9] A small ivory cross is shown in Figure 54.

Diptychs, Triptychs, Polyptychs, Plaques The writing-tablet aspect of the early diptychs is treated more fully in a separate section (see Writing Tablets, p. 193). A diptych consisted of two small rectangular tablets hinged together at one edge (Greek: *di,* two; *ptych,* fold); the two outer surfaces were often decorated with designs or various scenes of interest. The two inner surfaces of wooden or ivory diptychs were shallowed out and had a thin layer of wax on which temporary memoranda could be inscribed with a stylus and then erased by rubbing the blunt end of the stylus across the inscribed writing.

By the time that ivory diptychs were being carved for special use in Christian churches, the writing provision had become secondary (as had also happened with the secular diptychs). Then in later times the waxed writing surfaces were not provided at all; total emphasis was now placed upon providing a two-section story-device. For ecclesiastical diptychs, the portrayals on the two outer surfaces showed biblical events.

When, centuries later, the destruction of Christian symbolic representations began, some church curators reversed the hinges so that the diptychs could be folded back, showing only the blank (previously inner) surfaces on the outside. They were then stored upright on the book shelves between ordinary books in the church library. Whether this stratagem actually saved many ivory ecclesiastical diptychs from detection and destruction is uncertain. Cynics have suggested that any saving factor may rather have been the layers of dust on the book shelves, discouraging any inspection by the invaders of church rooms. Some art historians have said that earlier diptychs were later turned inside-out by way of reversed hinges for a different reason—to protect the carved ivory sides from continual wear as the diptychs were being used and moved upon tables.

In later times, when waxed writing surfaces were no longer provided, ivory carvers also fashioned ball-shaped diptychs of an ecclesiastical nature. The inner flat surface of each half-sphere had an incised biblical scene, while the outer curved surface was left plain, or was decorated with a simple design, sometimes appearing to be only an ivory carving of a fruit such as a melon.[10]

Eventually the diptych evolved into the triptych (three sections) and polyptych (usually five sections). In the three-section carving, the center part was usually twice as wide as each of the two end-sections. Thus, the left and right sections could be swung inward in front of the large center part, forming a compact closed work. There were also some five-section ivory carvings, consisting of two pairs of narrow, hinged-together sections, with one pair hinged on each side of a large single section in the center. The ivory carvings with several sections—three or five parts showing biblical scenes—were usually used as altarpieces and stood upright on a base. Some, however, were commissioned for use or display in the homes of affluent patrons of the arts.

Examples of the hinged ecclesiastical art-forms can be seen in London's British Museum and the Victoria and Albert Museum, as well as in Paris' Louvre and other major museums in Europe.[11] Sadly, one of the world's most magnificent ivory triptychs is marred by the fact that two pictorial panels are missing: one from the several small panels that were part of the left section and one from the several small panels of the right section. The missing panels may have fallen out and become lost, or may have been removed and stolen from a stored altar triptych (as has often happened with multi-piece carvings in churches and museums). This triptych is Florentine, from the sixteenth century, and is now in the Louvre.[12]

The Victoria and Albert Museum has an unusual example of how a special form of writing-tablet diptych was later converted into an illustrated biblical storybook. Originally, the two hinged tablets showed the customary carved biblical themes on the outer surfaces, but also served as book covers for six attached waxed writing tablets on the inside. The eight tablets are made of ivory. At some later time the wax was removed from the inside surfaces, and these were then colorfully painted with New Testament scenes.[13]

Single plaques or panels for biblical storytelling were sometimes set in a framed series within the church. The purpose was the same as the general function of the hinged works described above, but this kind of arrangement provided for a longer chronological series of biblical episodes. As a final note on the frequently mentioned storytelling function of religious art in church ivory carvings, we may wonder what proportion of the general populace actually got close enough to these carvings to see the portrayal of biblical figures and events.

Ecclesiastical Staffs A bishop's (or abbot's) staff is still carried as a symbol of office on ceremonial church occasions. It is more generally known as a pastoral staff or an ecclesiastical staff.

Because of its personal identification with the user, the staff was in earlier times sometimes buried with the deceased official, and a new staff was carved for his successor. Originally, they were made of wood and were of simple design. The top end was T shaped for the handgrip, thus suggesting the name tau (from the the Greek letter τ); the staff was also called a tau cross or a St. Anthony's cross. They are pictured in some eighth-century manuscripts. Of interest to us is the fact that they were occasionally made of ivory. Because of the precious nature of carved ivory staffs, only wooden staffs were deposited in the tombs of church officials. The earliest surviving staffs appear to date from about A.D. 1000. Many ivory ones were made during the eleventh and twelfth centuries.

Later, the simple T handle was replaced by a finely carved circular handle called a volute; the staff then became known as a crosier (or crozier). The volute—and sometimes a section added in the interior of the circle—would usually be carved with religious figures and scenes from the life of Christ. It may seem curious that some ivory volutes were in the form of a serpent; in those times, the serpent had a favorable symbolic significance, unrelated to the attitude of contempt shown toward that animal in the pre-Christian story of Adam and Eve.

One end of an ecclesiastical staff was sometimes hollowed out in order to hold the holy relics of some saintly deceased figure. This supplementary function of a staff is reminiscent of the secular canes that occasionally had a secret compartment (see Canes as Containers, p. 98).

Although old inventories in European churches frequently list ivory staffs, the handles are usually missing. This might be understandable in England, where most ecclesiastical ornaments were destroyed in the sixteenth century, but the same scarcity exists in France, Germany, and Italy.[14] Still, some complete ivory staffs have survived and are now in church treasuries or in museums. Examples can be seen in London's British Museum, the Victoria and Albert Museum, and in the Cluny Museum in Paris.[15]

Flabella Fly whisks made of ivory were most often used for preventing flies from desecrating religious objects, rather than for simply warding off annoying insects (see Fly Whisks, p. 158). Catholic flabella were especially used for driving away insects which sometimes alighted on or fell into the Eucharistic wine (and thereby symbolically desecrated Christ's blood). The standard flabellum was a fanlike whisk. The functional end, attached to a handle that was often made of ivory, was a square or circular shaped section of parchment or paper. The handles were sometimes exquisitely carved.

As Christianity spread over Europe, ivory flabella were made frequently, especially during the twelfth and thirteenth centuries; they remained popular until the fifteenth century. Examples, usually with only the handle surviving, can be seen in London's British Museum and the Victoria and Albert Museum, and also in the National Museum in Florence.[16]

The chowrie (or chaurie), made mainly in India, was different in construction but similar in purpose. It was used for brushing away insects from religious statues and articles. The ordinary chowrie consisted of a plain handle to which were attached a large number of slender strands. In some countries of the torrid zone, such whisks have been made of dried grass, vegetable fibers, feathers, or animal hairs. Some Indian chowries were made entirely of ivory. The "split ivory" chowrie required the painstaking and extremely detailed task of cutting a great number of very fine ivory strands down the length of a section of tusk until they reached the several inches of solid tusk that was reserved for the ivory handle. More simply, the strands were sometimes cut separately and were then attached to a carved ivory handle. A very attractive ivory chowrie is in the Victoria and Albert Museum; it is an eighteenth-century carving and is illustrated in a recent treatise on fans.[17]

Holy-Water Sprinklers Known as an aspergillum (or aspergill), this ecclesiastical device sometimes had an ivory handle. It was used at the asperges, a Catholic rite at which the holy water was sprinkled on the altar and clergy.

Incense Powder   Some ancient Chinese burned ivory powder as incense on altars and in front of sacred images.[18] Although powdered ivory might have been obtained specifically for such a purpose, it may rather have become available as the usual ivory-dust residue created by the sawing from the tusk of sections for ivory carvings.

Liturgical Combs   Though special ivory combs were used in certain church liturgies as early as the seventh century, little of definite nature seems to be known about the practice until much later times. By the fifteenth century, special regulations appear in the pontifical book of rites concerning ceremonial cleansing, including the combing of an officiant's hair. Pope Clement VIII (end of the sixteenth century) and Pope Urban VIII (early seventeenth century) strengthened and clarified the practice by issuing pontificals that listed the ecclesiastical objects necessary at the consecration of a bishop; included was an ivory comb. The custom of ecclesiastical hair-combing was gradually weakening, however, and was soon abandoned. Many of the specially made liturgical combs then found their way into the outside world.

Only a small number of the combs are known to have survived. Most of these are large, with a row of teeth on each side. Some have simple decoration with foliage and animals, representing a form of Christian symbolism of those times. Other combs have straightforward themes of a biblical nature. One or more can be seen in the British Museum in London, the Louvre in Paris, and the national museums in Brussels, Cologne, and Nuremberg, as well as in the church treasuries in some towns in France, such as the treasury of St. Etienne at Sens and the cathedral treasury at Nancy.[19]

Lockets for the Host   Similar to the European use of ecclesiastical pyxes for containing a supply of consecrated wafers (see Pyxes, p. 173) priests of the Eastern Orthodox Church once wore a small ivory locket containing a Host wafer when they visited a sick person who was unable to come to the church. Presumably the wafers were usually carried in a less impressive way. An example of one of the neck pendants, now in the Vatican Museum, has minutely carved scenes from the New Testament.[20]

Paxes   Apparently introduced in the thirteenth century, the pax (not to be confused with the *pyx*) replaced the earlier custom of personally exchanging a kiss of peace after the Mass. A pax, from the Latin for "peace," was a small devotional tablet bearing the figure or symbol of Christ or the Virgin Mary or a saint. It had a long handle for extending the tablet forward; choice tablets were made of ivory. In late medieval times, the priest kissed the pax and then sent the "kiss of peace" forward to the members of the congregation, who in turn kissed the pax. Thus it was appropriately called an *osculatorium*, literally, a "kissed object."

In Flanders, if members of a family were known to have quarreled irreconcilably, a pax could be sent there to bring about a reconciliation.[21] In Shakespeare's *Henry V* (act 3, scene 6), the unfortunate Bardolph is hanged for stealing a pax! Although paxes made of ivory were not very common, examples can be seen at the Victoria and Albert Museum in London, the Mayer Museum in Liverpool (England), and the Louvre in Paris.

Prayer Wheels Buddhist prayer wheels—prayer cylinders, actually—can still be seen in parts of India, Tibet, Nepal, Burma, and some nearby countries. At certain monasteries and temple sites, the cylinders were several feet long and were turned continuously by devoted followers or by a running stream (like a mill wheel). For individual use, prayer wheels were small devices that could be rotated on a handle for the purpose of repeatedly confirming the expression of prayers inscribed on the outside, and sometimes on the inside, of the cylinder. Each rotation constituted one implicit recitation of the prayers. Although most prayer wheels were made of wood and brass, some choice models were made of ivory. Figure 54 shows an ivory one bought in Nepal; the interior contains a long prayer-scroll.

Pyxes A pyx (Greek and Latin: *pyxis*, box) was a small cylindrical container made from the tusk of an elephant or walrus; the natural round shape was retained. It usually had a lid, similarly made of ivory. In Italy, these cylindrical boxes had already achieved popularity by the fifth century A.D.; Roman ladies used them as containers for jewelry and small personal articles. These secular pyxes often showed carved mythological figures or other representations. Their round shape is all that distinguishes them from many of the secular ivory containers discussed under the general category of Containers (see p. 95); but our interest here concerns their special adaptation to use by the Christian church.

The ecclesiastical pyxes were made primarily as containers for the consecrated wafers offered for the Eucharist communion, but some were used also for storing a small supply of the wafers before the consecration ritual. As might be expected, the carved representations usually showed scenes from the New Testament. It has been said that a substantial number of Christian pyxes were made during and after Charlemagne's revival of the Western Roman Empire.[22]

Reliquaries The use of very finely made cases, sometimes of ivory, to store the personal effects of deceased Christian figures who had achieved great merit and esteem appears to have begun around A.D. 300. The relics might also be bodily remains (bones and hair) or an object, or part of an object, associated with that figure. In later centuries, for example, many churches claimed to have a piece of Christ's last station on earth—the wooden cross. An early ivory reliquary, about A.D. 315, is at the Reliquary Museum in Brescia, Italy; the carved ivory sides show scenes from the Old and New Testaments.

Especially after Charlemagne's reign (about A.D. 800), as Christianity spread more widely over Western Europe, ivory was increasingly used for the outer covering of wooden and metal reliquaries. Because of the sacred nature of these containers, the carved ivory sides were often fine works of sculptured ecclesiastical art. In later times it would become a debated question as to whether all those treasure boxes actually contained authentic relics of the revered religious personages. A highly regarded reliquary, made of gold and ivory, is at the Herzog Anton Ulrich Museum in Braunschweig, Germany.[23]

One of the uses of oliphants (ivory horns) was as a special form of reliquary for very small sacred relics. An oliphant was made from the pointed end of the tusk; the solid interior was scraped out to make a short conical ivory container. In England, these oliphant reliquaries were sometimes hung in churches.

Retables   A retable (Latin: *retro,* back; *tabulum,* table) was a vertical partition that formed the back of an altar. When it was in the form of a stationary structure, it would usually be made of wood. When made of ivory, it was an ornamental screen that was brought out before the church services began. An ivory retable was basically an oversized triptych, the three large sections being hinged together so that this valuable altarpiece could be folded up for compact storage when not in use. Each large section consisted of small ivory panels, often carved elaborately with Christian themes.

Rosaries and Other Prayer Beads   Prayer beads predate their use in Roman Catholic worship; it is thought that originally a string of knots was used. Prayer beads have been made of a variety of materials, including clay, wood, glass, and metal. Non-Christian sets differ in the types of prayers recited, the number of beads included, and the use of differing sizes or colors among the beads within a set, according to the particular religion or sect. The use of ivory for non-Christian prayer beads is somewhat uncertain, but it is likely that a number of sets have been made of ivory.

Ivory has sometimes been used for carving the beads employed in the recitational counting of rosary prayers by Catholic worshipers. Many sets that appear to be ivory, however, are made of bone. The term "rosary" is derived from the Latin word *rosarium,* meaning "rose garden." Roses, symbolizing joy, were associated with Christ's mother, and especially with her divine motherhood and her later coronation as Queen of Heaven. She has sometimes been portrayed as wearing a crown of roses. The small beads, often fifty in number, provide for Hail Mary prayers; the large beads, usually one every ten small beads, provide for Our Father prayers.

Because the set of beads would often be carried on the person, many had connected ends so that the rosary could be worn around the neck, waist, or wrist. For the more expensive rosaries, the beads were carved in exquisite fashion. An impressive set can be seen at the Mayer Museum in Liverpool, England. The most lavish rosaries may have been made for display, rather than for meaningful worship. We note, for example, a sixteenth-century Flemish rosary in which each of six large and fifty-two small ivory beads showed four carved heads (back to back) of Christ and saints.[24] Finger rings, carved in the form of ten ivory rosary beads, are very rare now.

Shrines The Gothic period (mid-twelfth to early sixteenth century) furnished small Christian shrines and altars suitable for placing on a table. Some were made of ivory and were often self-enclosable within their hinged panels. Generally, they were intended for the bedroom as an accessory for private devotions. Their portability enabled travelers to carry one with them during periods away from church and home. Mexican priests sometimes carried a pocket-size model of an ivory altar.[25]

Japanese carvers similarly furnished small ivory shrines for table, desk, or shelf; an inner recess lay behind small ivory doors. Collectors will find them to be more readily available than the Christian models; the latter are quite rare now.

Talismans, Amulets, Charms, Fetishes The articles considered here represent a type of belief in nonmaterial forces or spirits which could bring good luck or deter bad luck or some evil. There is no sharp distinction between the meanings of the terms. Usually, however, a talisman is an object which is believed to attract favorable influences, while an amulet is one believed to ward off unfavorable influences. A charm, like a talisman, is generally positive in nature, as evidenced by the familiar good luck charms. A fetish, in some cultures, refers to an object in which a spirit is believed to reside and have magical powers favorable to the owner or carrier of the fetish. All of these articles have sometimes been made of ivory.

In ancient Egypt, until the twelfth dynasty (about 2000 B.C.), royal personages occasionally wore or carried an ivory amulet in the form of a magic wand, sometimes inscribed as being "for protection."[26] We may also note a very early ivory carving of a knot that appears to represent some obscure symbol; it is from Crete and is thought to be from about 1500 B.C. Similar representations of knots in Cretan seal impressions—as well as later representations of knots (in gold and alabaster) in ancient Greece—are believed to have been symbols used in religious rites.[27] The early Chinese made talismans and amulets from narwhal tusk; similarly, many centuries later, small talisman carvings from narwhal tusk were made by the Vikings.[28] Ivory good luck pieces became very popular in Europe between the mid-twelfth century and the early sixteenth century; they were especially valued when carried along on voyages and long trips, or any other occasion which entailed risk and danger. In Siam, ivory amulets were worn for protection from disease and other misfortune.[29] In Madras, India, some children wore special talismans made of ivory.[30]

Eskimos wore ivory charms to influence a hunted animal in a way that would be favorable to the hunter. In whale-hunting regions, for example, whale-shaped ivory charms were supposed to bring those animals into the regional waters. It is said that during a whaling ceremony (presumably before the hunting expedition) the whalers sometimes wore an ivory chain of whale-shaped links. Among Eskimos, ivory dolls were used in fertility rites, either to induce the birth of a son (preferably) or, more generally, to terminate a woman's condition of barrenness.[31] Similarly, in Cameroon (West Africa), women wishing to bear a child were sometimes given an ivory figurine of a pregnant woman, which would be worn as a talisman.[32]

Ivory pieces of the main kinds already considered—as well as fetish objects—were worn widely in Africa until the twentieth century.[33] But one of the most unusual ivory carvings ever to come to the attention of this writer was an Eskimo artifact: a walrus-ivory spirit catcher in the form of a carved tube.[34] We may wonder how the Eskimo managed to lure the spirit into the hollow trap and keep it there.

Vessels for Holy Water or Wine For ecclesiastical rituals and ceremonies, water and wine had sacramental value that was worthy of the use of ivory as the container. Especially from the time of Charlemagne's reign (A.D. 768–814), ivory vessels for water and wine were frequently used by the Church.

In Italy, the treasury of Milan's cathedral has a splendid ivory bucket for holy water, an example of Lombard artwork. This vessel is thought to have been made about 975.[35] A nineteenth-century reproduction is at the Fowler

Museum of Decorative Arts in Los Angeles. London's Victoria and Albert Museum has a fine ivory "Holy Water Vat," made for the Milan ceremonies of 983, the year of Otto III's succession to the title of Holy Roman Emperor (although, at three years of age, a regent ruled in his place). The exterior ivory carvings show Christ washing the feet of his disciples and other scenes from his later life.[36] Church entrance stoups, basins for dipping the fingers before blessing oneself, were sometimes made of ivory. An early German ivory stoup is thought to date back to the tenth century.[37]

In the Monastery of Xeropota at Mt. Athos in Greece, there is an ivory cup that was carved with figures of the Virgin and Child surrounded by apostles and prophets. The monks claimed that water drunk from that cup would counteract poison; thus, we find in religion too the common belief that ivory can serve as an antidote to poison. More curiously, the monks refused to let the cup be photographed, lest its miraculous curative power be lost.[38]

18 ⊛ SCIENTIFIC INSTRUMENTS AND AIDS

Scientific instruments and aids—if defined broadly enough to include the measurement, calculation, and observation of natural phenomena—employed ivory in various ways. In some instances, as might be expected, ivory could not serve as the functional part of the article.

Architectural Compasses The hinged two-pronged compass made of ivory began to appear in India around the sixteenth century. The Colombo Museum in Sri Lanka has a seventeenth-century model with ivory prongs and a metal hinge.[1]

Calculators Sixteenth-century woodcut prints show that a simple form of reckoning board was made of ivory.[2] A kind of pocket calculator made of ivory was popular in Western Europe in the seventeenth and eighteenth centuries; the device was known as "Napier's bones." John Napier (1550–1617) was a Scottish mathematician who invented the logarithm system. His calculator had an arrangement of numerically marked rods, or "bones," that facilitated the arithmetical operations of multiplication and division. An ivory model made around 1700 is in the Museum of the History of Science in Oxford, England.[3]

Early slide rules used wood, brass, or ivory for the scales. Ivory was especially favored because the contrasting black numerical markings on the whitish background made the readings highly visible. In the 1880s, Germany

[178] Part Two : Uses of Ivory

began to furnish these calculators with scales made of pure white Celluloid, offering even better contrast.

Conversion Scales Before the advent of Celluloid, various kinds of testing equipment had an attached arithmetical conversion scale that was often made of ivory. It looked like a ruler with one set of numerical readings across the top edge and a different set across the lower edge. Unlike the slide rule, it had no moving parts and, more importantly, it could only be used with the particular equipment for which it was made. For example, a certain instrument was designed to measure the specific gravity of a liquid sample of alcoholic beverage; an attached ivory scale also showed the percentage alcoholic content equivalent to the specific gravity reading.

Directional Compasses An ivory band sometimes served as the rim of small hand-held compasses, indicating north, south, east, and west. They are not easily found by collectors of utilitarian ivory articles.

Navigational Instruments Hippopotamus ivory was once used for the small movable auxiliary scale on sextants.[4] Some sixteenth- and seventeenth-century navigational instruments, such as astrolabes, backstaffs, and cross-staffs, were made entirely of ivory, with black engraving for the markings.[5]

Planetariums The first synchronized mechanical model of the solar system was apparently the one constructed in England around 1709. Then, about 1712, John Rowley made one for Charles Boyle, the Earl of Orrery, and named it an "orrery" after him. One model is in the National Maritime Museum in Greenwich, England; brass rods hold small ivory spheres representing the planets and several satellites.[6]

Telescopes and Binoculars The tubular metal framework of earlier telescopes and binoculars sometimes had a veneer covering of ivory or bone. Collectors may occasionally obtain a bone-covered pair of binoculars in an antique shop or pawnshop; most that may appear to use bone or ivory, however, actually employ Celluloid or similar white plastics.

19 ⊛ SMOKING ARTICLES AND ACCESSORIES

The use of ivory for smoking articles has been almost entirely limited to serving as the material for containers and holders. Among the few exceptions are cigar-tip cutters, pipe stoppers, and tobacco graters.

Cigarette Cases and Cigar Boxes As with most other containers, the terms "case" and "box" are often used interchangeably in connection with cigarettes and cigars. Figure 56 includes an ivory pocket-model cigarette case, probably made in India; the cover is painted with a tiger and an Indian elephant. Also shown is an ivory cigarette case for table or desk, from Persia; the top and sides are colorfully painted. The other container is a cigar box from Alaska; the cover of the wooden box is ornamented with ivory carvings of an Eskimo and his sled and dog team.

Cigarette Holders and Cigar Holders Large numbers of cigarette holders have been made of ivory in various countries, especially China. The Chinese examples were often carved with figures of dragons; some have been signed. Figure 57 includes one of these, together with the following: a cigar holder, carved with a bird and foliage; a cigarette holder with an ivory mouthpiece; a small ivory cigarette-holder with its ivory case, both with painted scenes; and a small collapsible ivory and brass cigarette-holder with its tiny cloisonné container (the metal link makes it possible to attach the container to a bracelet or other costume article).

Ivory collectors have little difficulty in obtaining a cigarette holder; some are still being made. Cigar holders made of ivory are much more difficult to find.

Cigar-Tip Cutters To avoid loss of flavor before use (especially until the advent of airtight sealed wrappers), many brands of early cigars had a tip that was sealed with tobacco leaf. Some brands, including the finest kinds, are still made that way. For cigars that were not made with a pre-cut tip, the end had to be cut off or cut through. A special device had a steel cutting-element; some of the more expensive models were mounted on an ivory handle. Figure 57 includes two early examples of cigar-tip cutters. For one, a small tusk serves as the handle; it shows the cracked appearance of some old ivories. For the other, the ivory handle is a carved branch with a crawling snail.

Hookah Stands During the nineteenth century, India was furnishing ivory-inlaid hookah stands. A hookah is a smoking pipe with a long flexible tube leading from an urn of water. The smoke is cooled by passing through the water before it is drawn into the mouth.

Match Safes These small cases provided a portable container for matches carried in a pocket, affording protection against fire if match heads ignited by striking against each other. Fine little cases, often made of sterling silver or ivory, were popular from the middle of the nineteenth century until the early part of the twentieth century. Then they were replaced by matchbooks of safety matches, as well as by cigarette lighters. An ivory model is shown in Figure 57; one outer edge is ribbed so as to serve as a convenient striking surface.

Pipes, Pipe Stoppers, Pipe Cases Unless a more resistant material lines the tobacco bowl of a pipe, the burning tobacco will scorch the ivory and may emit an unpleasant odor. Solid ivory pipes without any liner, however, have been made. Alaskan Eskimos made them from walrus ivory.[1] Some Eskimo pipes are at the University of Pennsylvania Museum in Philadelphia;[2] examples can also be seen at the American Museum of Natural History in New York City.[3]

As an accessory device, Victorian England furnished ivory pipe-stoppers (tampers) for stuffing the tobacco down into the bowl. These were often elaborately carved, representing Victorian celebrities or animal grotesques.[4] American seamen carved them from whalebone; some of these had an extension at the opposite end of the tamper which served as a pick for pushing air holes down to the stem if the tobacco had been pressed down too tightly.[5]

A variant of the ordinary pipe was the opium pipe. Carved from ivory or bone, it resembled a flute; many were about 10 inches (25 cm.) long and $\frac{3}{4}$ inch (2 cm.) in diameter. China produced a large number of them during the eighteenth and nineteenth centuries. Then, shortly after the end of the last dynasty, in the early twentieth century, China outlawed the growing of the opium poppy.[6] A small ivory box for opium powder and an ivory spatula accessory can be seen in the Chinese exhibit at the Field Museum of Natural History in Chicago. An opium pipe is at the Fowler Museum of Decorative Arts in Los Angeles.

Most pipe cases made of ivory have come from China and Japan. A Chinese pipe case from the Chien Lung period (1735–1796)—decorated with a moon, cloud, bat, and plant—is at New York's Metroplitan Museum of Art.[7]

Tobacco Boxes and Jars In addition to small portable tobacco boxes made of ivory, larger boxes or jars were made for keeping on a shelf or desk. During the seventeenth and eighteenth centuries, ivory tobacco-jars became very popular in Western Europe. China furnished smaller ivory jars for opium powder.[8]

Tobacco Graters Pre-cut tobacco for pipes and shredded tobacco for hand-rolled cigarettes have long been commercially available. However, tobacco graters were necessary in the early days, and it is possible that some were made of bone or ivory. When taking snuff first became popular, pulverized tobacco for sniffing also had to be prepared by the users themselves. (Since snuff graters are not a smoking accessory, they are discussed separately, on p. 200.)

20 ❀ TOYS

As distinct from the numerous different kinds of ivory games and puzzles (see pp. 122, 165), only a small variety of toys made of ivory have appeared. One reason for this may be that the cost of ivory has precluded any prolific use for children's toys. Most ivory games and puzzles were intended primarily for adults.

A very ancient and clever ivory toy was a set of four small carved ivory figures of dwarfs or pygmies, which could be operated to make the figures perform a sort of dance. It was unearthed in the 1933–1934 excavations of a girl's tomb in Egypt dating from the twelfth dynasty (1991–1786 B.C.). In fact, three of the four figures were actually dancers, on spools that could be rotated manually by strings running through each figure's base (the strings were, of course, no longer present). The American excavation team retained the fourth figure—which apparently provided the rhythmic beat by clapping its hands together—for New York's Metropolitan Museum of Art; the three dancing figures were presented to the Egyptian Museum in Cairo.[1]

It was not until fairly recently, in the nineteenth century, that ivory toys began to be made in any significant number. Then the American whaler-seamen started to make toys from sperm-whale teeth, whalebone, and baleen. For girls, they carved dolls and small dollhouse furniture and furnishings (including tiny clothespins for hanging up the doll laundry to be dried). Boys were given spin tops, puzzle boxes, and a variety of miniature replicas of weapons—knives, clubs, harpoons, and so on.[2] Ivory or bone toys are now extremely difficult for collectors to obtain.

Ball-and-Cup Toy Made in many countries, usually from wood, this toy has also been carved from whalebone (and possibly ivory) by American whaler-seamen. The top of the handle was cup-shaped and had a string attached to a small solid ball. The ball had to be swung outward and upward, and then immediately caught in the cup as the ball fell down.[3]

Dolls Perhaps the earliest ivory toy was made by a caveman father who etched the figure of a doll on a piece of tusk for his young daughter. But few ivory dolls seem to have been made during the many centuries following. Sometimes an antique dealer or a doll repair shop will report having seen a small ivory or bone doll from the eighteenth or nineteenth century. Small ivory dolls carved from walrus tusk by Alaskan Eskimos can be seen at the United States National Museum.[4]

Fortune-telling Devices Unlike many ancient methods of divining the future, such as throwing yarrow stalks, Figure 53 shows a nineteenth-century decision-maker requiring a very simple approach. Obtained in Persia, it is in the form of a small bone or ivory fish comprised of seven sections strung together like beads on a string; the string extends about 1 inch (2.5 cm.) beyond the mouth of the fish. To make a well-fated decision (yes or no), you hold the supporting string between the thumb and forefinger of one hand, behind your back (or in front, but with eyes closed). The other thumb and forefinger blindly grasp a section of the fish, while announcing either choice (for example, no). Then, those two fingers move upward along the fish sections, each time with the announcement of the opposite to the one preceding it. The last section grasped (the fish head) determines the right decision.

Rattles Alaskan curio dealers have reported that Alaskan Eskimos sometimes made an infant's rattle from walrus ivory, but the present writer has not seen one, nor a photograph of one. A combined rattle and whistle made of ivory and whalebone by a scrimshaw carver is at the Mystic Seaport Museum in Mystic, Connecticut.

Tops In the Greek comedy *The Birds* (written in 414 B.C.), Aristophanes mentions a whipping top. Some early Greek tops were probably made of ivory, for not too long afterward, early Roman tops for gambling were made of ivory and had flat sides imprinted with the various "put" and "take" designations (see Tops for Gambling, p. 129). Alaskan Eskimos carved ivory tops that had a vertical peg penetrating a flat tutu-like disc.[5] Humming and whistling tops have been made by cutting grooves or holes in the top, so that air rushes through when the top is spinning. Japan made metal tops of this kind, but India made some from ivory.[6]

Whistles A small toy whistle made of ivory is shown in Figure 53. Larger whistles of ivory or whalebone are seen occasionally. An ivory one can be seen at the Santa Barbara Historical Society Museum in California.

21 ✦ WEAPONS, HUNTING AND FISHING ARTICLES, AND ACCESSORIES

It may seem surprising that ivory has been used not only for accessories in hunting, fishing, and fighting, but also directly for the functional part of some weapons of attack or capture. Examples include arrowheads, fishhooks, harpoons, spear points, swords, and bolas.

Archers' Thumb-Guards and Wrist-Guards Among ancient Chinese, when the taut bowstring was pulled back and released by the thumb (instead of the three middle fingers), a ringlike guard, often made of ivory, protected the thumb from the sudden harsh scraping action of the bowstring. More recently, in India, ivory thumb-guards were made in the fifteenth century.

Also sometimes made of ivory (but usually of leather) were the wrist-guards worn on the arm that held the bow. When the bowstring was released, it might scrape or strike the inside of the forearm, inflicting painful injury if no protection was afforded by a wrist-guard or arm-guard. The Wallace Collection in London has a wrist-guard richly carved in ivory.[1]

Arrowheads and Spear Points For arrows and spears, sharpened ivory points have sometimes served as the penetration agent when metal was unavailable. Ivory and bone arrowheads were found during excavations in Egypt of predynastic tombs (dating from before 3100 B.C.). In later ages, Eskimos used ivory arrowheads for killing bears.

The writer has seen only a single reference to spear points made of ivory.[2] A personal communication from that source indicated that they had been made in northern regions as far apart as Alaska and Greenland.

Arrow-Shaft Straighteners For shaping an arrow, straighteners were made of bone or ivory. They looked like a narrow flat stone with one or more holes—the diameter of an arrow—passing through the flat sides. Holes were often made at both ends of the straightener; the end opposite the hole being used would serve as a handle. By passing a rough wooden shaft through the hole, a hunter could rub the shaft down until it was smooth and of uniform thickness; the arrow point was then installed at one end.[3] Figure 58(f) shows a shaft straightener made from bone. The two small holes may have had some other purpose; the large hole at the opposite end seems to have been worn into an ellipse through constant use.

Body-Armor Plating As early as A.D. 262, the Su-shen tribesmen of northeast Asia were making armor out of rows of small overlapping walrus-ivory platelets. These were much lighter than iron armor and more enduring than leather armor.[4] Many centuries later, Eskimos made the same product, lashing the walrus-ivory discs together with rawhide strips through holes in the discs.[5] Figure 58(e) shows a rectangular armor platelet made from a seal rib-bone by Alaskan Eskimos in an earlier time. Ivory armor is extremely rare now.

Bolas For striking down small game and edible birds, Eskimos threw a walrus-ivory bola. It consisted of several ivory pieces, each attached to the end of a separate cord, with all the cords attached together at the opposite end.

Bow-Tips Ornamentation In early China, ivory tips sometimes decorated the ends of a warrior's bow.[6] In recent centuries, entire bows were often made of baleen, the flexible strips that hang down inside the mouth of the baleen whale.[7]

Daggers and Swords The use of ivory for swords and daggers has been limited to the handle and the hand-guard. The British Museum in London has two Egyptian daggers with inlaid ivory ornamentation in the handles; they date back to about 1800 B.C. Some time between about 1500 and 1000 B.C., China's first dynasty was furnishing ivory guards for swords.[8] About two thousand years later, from the eighth to twelfth century, the Vikings carved entire swords from the narwhal's long straight tusk.[9]

It was once believed that ivory weapons had other virtues too. In an earlier section on walrus ivory in Part One (see p. 31) it was mentioned that some Scandinavian warriors believed that ivory handles on knives, swords, and javelins could magically ward off cramps. Similarly, Moslems once thought that ivory guards on daggers could stanch the flow of blood as well as induce faster healing of wounds.

England produced many ivory handles for daggers in the twelfth and thirteenth centuries. An ivory sword-handle from France, dated about 1350, can be seen at London's Victoria and Albert Museum. This handle illustrates the decorative extravagance that ivory carvers often lavished upon such objects.[10] At one time, Ceylonese carvers made ivory hand-guards for fencing foils. More recently, nineteenth-century American whaler-seamen carved dagger handles from ivory and whalebone.[11]

Japan has furnished highly decorative ceremonial daggers with the handle and entire sheath made of ivory. They are often pictorially incised and colorfully painted, and have an attractive tassel attached. One is shown in

Figure 55. In the United States, they are sometimes mistakenly called hara-kiri daggers (Japanese: *hara,* belly; *kiri,* cut), after the samurai practice of performing suicidal disembowelment with a sword when suffering great personal disgrace. Also shown in Figure 55 is a dagger with a carved ivory handle and a sheath with a decorative ivory mouth. This dagger was bought in the Philippines, though its origin is uncertain.

Fishhooks Ivory fishhooks were found in Egyptian predynastic tombs (before 3100 B.C.). More than four thousand years later, walrus-ivory fishhooks were being carved by Eskimos.[12] Figure 58(c) shows one from Alaska with small piece of the shank broken away.

Fishing Lures Figure 58(d) shows a walrus-ivory fishing lure (without hook), carved and painted as a small black fish; it was made by Alaskan Eskimos.[13]

Harpoons Whereas a land animal could be hit with an arrow or spear and then followed or tracked until it collapsed, a wounded sea animal usually submerged and could therefore only be followed at the end of a line. A hand-thrown harpoon consisted of a shaft with a trailing cord attached to a sharp harpoon head at the front of the shaft. The weapon was hurled like a spear. As soon as the large fish was hooked by the detachable harpoon head, the head came off the shaft; the thrower would then use this hook-and-line device to pull in his catch after the fish was exhausted.

Harpoons were already being used by ancient Eskimos, who decoratively engraved some ivory harpoon-heads. Examples of ivory harpoon-heads can be seen at the University of Alaska Museum in Fairbanks [14] and also at the U.S. National Museum in Washington, D.C.[15] Figure 58(b) shows an Alaskan Eskimo carving said to be a small harpoon head; one of the two rear barb points has broken away.

Harpoon rests and spear rests were sometimes made of ivory. Harpoon rests carved by Alaskan Eskimos were often embellished with representations of whales. Examples can be seen at the U.S. National Museum and at London's British Museum.[16]

Hunting Horns Like ivory horns used as musical instruments, ivory hunting horns were made by hollowing out the solid pointed end of a tusk (see Horns and Trumpets, p. 137). A lavishly carved Arab horn made in Sicily as early as the eighth century is at the Ulrich Museum in Brunswick, Germany.[17] A German hunting horn supposed to have belonged to Charlemagne (A.D. 742–814) is at the Schatzkammer Museum in Aachen,

Germany.[18] London's Victoria and Albert Museum has a magnificent fifteenth-century German horn carved all around with a multitude of finely detailed Christian events, hunting scenes, romantic scenes, foliage, and fabled animals.[19]

Pistols and Rifles Starting in the fifteenth century, various kinds of firearms were often inlaid or veneered with sections of ivory. The following century introduced a practical form of pistol; before long, the handles of choice models sometimes had solid ivory panels for the side-grips. The Wallace Collection in London has a fine group of artistically embellished firearms and other weapons. The decorative representations are quite varied; they include hunting scenes, mythological and allegorical figures, real and imaginary animals, and monograms.

Powder Horns After the invention of gunpowder, portable powder flasks were made from either animal horn or a hollowed end-section of a small tusk. The pointed tip was cut off, and a stopper was inserted; when opened, gunpowder could be easily poured into the firearm. A cap covered the wide end and could be removed for filling the horn. Such a container kept the powder dry, and the ivory models were often embellished with carved designs. Powder horns made of ivory were crafted widely in Europe, with many notable ones coming from Germany; some were also carved in India. Finely carved European models are in the Wallace Collection in London.[20] Several ivory powder-horns are at the Fowler Museum of Decorative Arts in Los Angeles.

Seal Attracters To attract seals, some Eskimo hunter of long ago had the ingenious notion of scratching on the ice with a section of a dead seal's foot and claws in order to duplicate the sound that seals hear when a seal mounts the slippery ice floes above. A later form of this lure eliminated the need for the actual foot by substituting a handle tipped with only the claws. The final development did not have the natural claws; it used a scratcher shaped as claws. Figure 58(a) shows a small ivory model.

Shields Greek mythology credited Heracles with a shield ornamented with ivory, described by Hesiod in the eighth century B.C. in his *The Shield of Heracles*. Actual ivory shields, presumably made of pieced sections, are said to have been used during the Roman Empire.[21] We may wonder whether any of those shields have survived.

Truncheons Nineteenth-century American whaler-seamen carved truncheons (billy clubs) from whalebone.²² One containing a solid ivory whale tooth on the end of its handle can be seen at the Mystic Seaport Museum in Mystic, Connecticut.

Whips An ivory whip with detailed hieroglyphic inscriptions was found in the fourteenth-century B.C. tomb of the Egyptian king Tutankhamen.²³ It may seem surprising that ivory could be used for a whip, but until ivory dries out after many years, a slender length retains a substantial degree of flexibility. It is uncertain as to whether other ivory whips have survived. During the nineteenth century, American whaler-seamen sometimes cut a whip from baleen.

22 ✦ WRITING AND PAINTING INSTRUMENTS, AND RELATED ARTICLES

Writing and painting instruments, as well as articles either directly or indirectly related to them, have employed ivory in a great variety of ways. As with most other types of ivory products, ivory itself does not suggest any obvious uses; the possible applications have been discovered through the imagination and ingenuity of craftsmen.

Architect's Ink-Pens Figure 59 includes an architect's ink-pen with an ivory handle; these were made in several sizes.

Burnishers Howard Carter, discoverer of the fourteenth-century B.C. tomb of Egypt's King Tutankhamen, tells about one article found there: "The use of the elegant but curious mallet-like ivory instrument is not so easily recognized; however its gold-capped top suggests it is a burnisher, for smoothing the rough surfaces of the papyrus paper."[1] The head is skillfully crafted and has painted inscriptions. A recent source says the head's flat top (the rubbing surface) is covered with gold foil.[2]

Business Cards This writer knows of only one instance of an ivory business card. Paul Revere (1735–1818)—American patriot, silversmith, and engraver—once engraved an ivory business card for a Boston ivory turner.[3] One can only wonder where that ivory card is now. A single expensive card of this kind must of course have been retrieved each time it was presented. (See also Nameplates, p. 189).

Dance-Card Booklets Similar to the dance fan (see p. 140), this small booklet of ivory cards enabled a popular young lady to write down the names of the gentlemen whose request for a dance had been granted. Afterward, the penciled names were easily erased from the ivory surfaces (Fig. 59).

Hand Rests Especially during China's last dynasty (Ching, 1644–1911), ivory hand-rests were a cherished possession. They were not for resting a tired hand, but rather for steadying the writer's wrist or lower arm while writing small characters. About ½ inch (1 cm.) high, the supports were originally made of bamboo. Ivory rests from the Chien Lung period (1735–1796) can be seen in the Chinese exhibit at Chicago's Field Museum of Natural History. Relief carving often showed a complete landscape.[4]

India Ink and Ivory Black Residual ivory dust from the sawing and filing operations has had several uses. A major use came after the discovery that ivory dust, it burned under controlled conditions, could then be processed to make india ink, as well as artists' oil color known as ivory black. Eventually, however, lampblack replaced ivory dust for making india ink; and burnt leather replaced ivory dust for making black oil paint. Interestingly, ivory was once used for making another color: in Alaska decayed mammoth ivory of bluish color was ground up by Eskimo artists and was then used as blue pigment for decorating the souvenirs they crafted.

Ink-Blotter Holders Ink-blotter holders were often made with a curved bottom so that the attached blotter could be rolled across the paper's wet ink surface. These devices, whether made of ivory or cheap materials, provided for replacement with a new blotter. Some of the ivory models were made in India, especially during the nineteenth century. Figure 60 shows a solid ivory model.

Ink Scrapers These ink eradicators merely scraped away the dried ink. Figure 59 shows one with a double edge and a handle made of ivory or bone, manufactured in Sheffield, England.

Inkwells Excavation of the Assyrian ivories at Nimrud in Iraq (from the period 1200–600 B.C.) yielded an apparent inkwell formed by two bearded male sphinxes.[5] Until the twentieth century, ivory inkwells were often crafted in the Moslem world. A substantial number came from India during the nineteenth century. At that time, American whaler-seamen made inkwells of ivory from sperm-whale teeth.[6]

Labels Egyptian tombs of the kings from the first dynasty (3100–2900

B.C.) contained small ivory labels on jars of balsam-tree oil. These and other ivory labels had a hole so that they could be attached to jars and other objects to indicate ownership. A king's name was incised on each label.[7]

Letter Openers These items are also known as paper cutters and paper knives; ivory models have been made in many countries. Although some were quite simple, produced on machines, a large number were hand-crafted in interesting shapes and designs, especially of animal figures. A popular model from India shows a line of elephant figures along one edge of the cutter. Figure 52 includes several forms of the letter openers, including one from China with a steel blade and ivory sheath.

Memorandum Tablets and Pads Only memorandum tablets and pads with ivory writing surfaces are discussed here; waxed tablets are considered separately (see Writing Tablets, p. 193).

By the time that the Chou dynasty ended in the third century B.C., important Chinese personages sometimes carried an ivory writing tablet *(hu)* suspended from the waist girdle. Later, in the Tang dynasty (A.D. 618–906), these tablets could be used only by officials of the six highest ranks. Still later, during the Ming dynasty (1368–1644), the use of the tablets were further restricted to the four highest ranks.[8]

Examples of such ivory tablets can be seen in the Chinese exhibit at Chicago's Field Museum of Natural History. These tablets are about 4 inches (10 cm.) wide and about 15 to 20 inches (38–51 cm.) long. The 4-inch width at one end tapers down to a narrower width at the other end. The memorandum tablets were used for recording a transaction with an esteemed official of the court. We are informed that the broad end of the tablet was held up in front of the speaker's mouth so that he would not breathe upon the esteemed official's face.

During some periods in China, the shape and size of the ivory tablet indicated the rank of the official. If he addressed the emperor, he would have his speech written in advance on the ivory surface.[9]

Memorandum pads have been crafted in recent years, consisting of several very small ivory sheets on which a temporary note could be written in pencil and then erased. Figure 59 shows two kinds. One is a six-day diary, each sheet headed with captions from Monday to Saturday; in a side slot, there is a tiny pencil with an ivory cap. The other is a dance-card booklet (see p. 187).

Nameplates The ivory models were small rectangular strips on which the name of a person or agency could be engraved or painted; they were attached near the doorbell or at eye level on the entrance door. In offices, individuals sometimes mounted an ivory nameplate on a small wooden block on their desk. In the United States, the Pratt-Read Company in Connecticut produced such machine-cut strips ready for inscription. Figure 59 shows an old nameplate from England, as well as a recent one (incised on old ivory) for this writer.

Painting and Writing Brushes The Chinese used brushes that sometimes had a handle made of ivory; when not in use, the brush might be kept in a slender ivory case. An example of a brush-and-case set can be seen in the Chinese exhibit at the Field Museum of Natural History in Chicago. In the artist-scholar's hand, the same brush could either be a painting instrument or a writing instrument; the writing aspect was the more highly honored art.[10] Some ivory models have been fondly engraved with a poem or cherished classical quotation, perhaps with a little floral decoration too, all minutely executed by hand. If the brush were a personal gift to someone, the donor's name and the honored recipient's name would sometimes be added.

Paintbrush Cups Paintbrush cups or holders made of ivory were carved in China during the last dynasty (Ching, 1644–1911).[11] Examples can be seen in the Chinese exhibit at the Field Museum of Natural History in Chicago; they were cut from a short round section of the tusk. Ivory rests for paintbrushes are sometimes mentioned.

Paintbrush Palettes In the early part of the twentieth century in London, the Hilton Price Collection of Egyptian Antiquities had a scribe's palette made of ivory, $13\frac{1}{8}$ by $1\frac{3}{8}$ inches (33.5 × 3.5 cm.); it is thought to date from the eighteenth dynasty (1567–1320 B.C.). Two cavities at one end had once contained red and black paints; a longitudinal groove along the center had housed the reed pens.[12] Howard Carter, discoverer of the fourteenth-century B.C. tomb of Egypt's King Tutankhamen, found several ancient ivory palettes, one with six separate color cavities (the dried paints still visible).[13] New York's Metropolitan Museum of Art obtained such a palette, inscribed with the name of Amenhotep, presumably belonging to one of the Egyptian kings bearing that name in the fourteenth century B.C.

Another ivory artifact found in King Tutankhamen's tomb was an accessory for a palette—an artist's water bowl $6\frac{1}{2}$ inches (16.5 cm.) in diameter carved from a single large block of ivory. It is color stained from long use in ancient times.[14]

Paper Creasers Not to be confused with letter openers, paper creasers (also known as sheet folders) were flat rectangular strips of ivory, without the cutting edge or tapering width seen in letter openers. They were used for folding a large number of printed sheets into two or more pages or sections. Figure 52 includes one made in the United States by the Pratt-Read Company in Connecticut; at 10 inches (25 cm.) in length, it is rather longer than most. This company furnished them to U.S. Congressmen and called these paper creasers "Congress folders," a name and a product that have not endured.

Paper Rougheners A small rod-shaped object with a rough undersurface was once used to prepare a small area on a paper sheet before applying heated wax, so that the softened wax would stick tightly to the paper. Then an incised name-design was pressed upon the wax to register a seal impression. Also, before the invention of envelopes with gummed flaps, the heated wax was used for sealing a folded letter. Figure 59 includes two kinds of rougheners: the tiny one is solid ivory with a rough underside; the more typical, larger one has a rough brass plate at the bottom of the attractive ivory handle.

Paperweights It is probably not possible to establish that a certain ivory carving was initially intended to serve the humble function of holding down sheets of paper. Perhaps the ivory piece was originally a simple flat ivory carving which someone later utilized as a paperweight. Ivory carvers sometimes specified the paperweight function for some of the small bulky pieces they crafted (Fig. 60).

Pencils In addition to the miniature ivory pencils encased in some memorandum pads and calling-card cases, larger ones were crafted too. As with numerous other utilitarian ivory products, some were intended as fine souvenirs commemorating an important public event or a visit to a prominent site. Figure 59 includes an ivory-handled pencil with see-through magnified views in the carved hand at the top end—three pictures of Napoleon and one picture of his tomb in Paris.

Pen-and-Pencil Trays Although these rectangular trays were occasionally made entirely of ivory, the material was usually only used for inlays. In India, the basic wooden structure of the tray was sometimes ebony. Figure 59 includes a writing instrument tray from India with an ivory inlay of an Asian elephant; the surface of the wood is painted.

Pens and Pen Cases An Egyptian pen case of ivory was found at the site of ancient Megiddo in Palestine; it is thought to be from the period 1350 to 1150 B.C.[15] Produced about two thousand years later, an eleventh-century Anglo-Saxon pen case made of walrus ivory can now be seen at the British Museum in London. The case is $9\frac{1}{4}$ inches (23.5 cm.) long, and shows scenes of men hunting animals.[16]

Penholders on which a detachable nib was fitted were among a large variety of ivory articles made in India during the nineteenth century. Indeed, they were crafted widely among the countries that furnished ivory products. Alaskan Eskimos made them for the tourist and export trades. Figure 59 includes two such penholders with ivory handles. The larger one was made in Germany but was inscribed in England: "Brighton Fair 1914." The attached nibs are not the original ones.

Potters' Dies Small engraved design tools for impressing the design upon soft, newly fashioned pottery before the hardening process occurs were sometimes made of ivory. Potters' dies made of ivory were crafted in Ceylon (and perhaps elsewhere too).

Pounce Pots These were small containers, perforated at the top, containing ground fish-bone powder or other absorbent powder—even very fine sand—to be sprinkled upon wet ink writing. Ivory was seldom used for these, although American whaler-seamen made a number of them from whale-tooth ivory.

Seals, Seal Boxes, Seal-Pigment Boxes In China, ivory signature seals—for businessmen, officials, and other members of the privileged class—date back to the Han dynasty (206 B.C.–A.D. 221).[17] Considering the expensive nature of ivory, the question is sometimes asked as to why some Chinese seals (often those with a carved animal figure at the top) were substantially longer than necessary. The answer is apparently so that each following generation could grind down and eradicate the previous signature surface at the bottom of the seal and then have its own signature incised there; thus, the handle became shorter and shorter, from the bottom upward, but still retained the carving at the top. In an ascending scale of value and dignity, an ivory handle was considered to be of the highest level, with only one exception—jade handles, reserved for the royal family. Bright vermillion was the favored ink color for seal impressions.[18]

The use of the ivory seal in Japan was imitative of the Chinese practice. Some Japanese seals did not have an incised name, only a felicitous greeting, such as "Happiness and long life."[19]

In England, ivory seals are thought to date back to the thirteenth century, perhaps even earlier. During the nineteenth century, American whaler-seamen carved these signature devices from whale-tooth ivory. Early ivory seals can be seen in London's British Museum.

Figure 60 shows a variety of ivory seals. The largest one has been carved minutely with floral designs; the two ball-shaped sections have been perforated. The machine-turned ivory seal-handle has a brass seal showing a bust of Beethoven, perhaps sold as a souvenir at a commemorative event, such as one of the 1870-centennial-year celebrations of the composer's birth. The handle with lovers' hands enclasped has a silver seal with a royal crown design; this may have been a family seal for husband and wife of some eighteenth- or nineteenth-century monarchy. The short bulky seal is made entirely of ivory, identifying a company in Rangoon, Burma. The tiny seal, made in Japan for an Englishwoman, is of tortoise shell with an ivory inset reading "Henrietta" (in Japanese characters).

In China, special boxes for ivory seals were crafted during the Sung dynasty (A.D. 960–1279); some of these were made of ivory. Also furnished were small ivory boxes for the color substance. Later, the portable Japanese inro was sometimes carved exquisitely from ivory, though it was, more commonly, a highly skilled form of lacquer art. They had small separate compartments for seals, seal-pigment containers, and other articles. Figure 60 includes a small ivory box for seal pigment; it was probably made in China.

Snow Knives Ivory snow-knives were similar to letter openers, but somewhat more solid. Natives of the far north of America used them to carve figures and simple illustrative scenes out of the Arctic snow when telling a story to children.[20]

Water Spoons In India, a special spoon made of ivory was used for adding water to ink, though it is likely that any small spoon would have served as well.[21]

Writing Styli and Tablets The excavations of Assyrian artifacts, from the period of about 1200 to 600 B.C., yielded ivory writing styli. At some time in ancient history, pointed styli began to be used for inscribing written characters in a soft substance such as clay or a waxed surface.

One of the earliest devices for a portable and erasable writing surface was the waxed-surface tablet, which was square or rectangular in form. The surface was ground down slightly within the four edges and was then filled with a thin layer of wax. The inscribing was done with the pointed end of the stylus; its opposite end was flat and thus served as an erasing head when it was rubbed over the inscribed impressions. These earliest tablets were made of wood. An improved form had two hinged tablets with the waxed writing surfaces protected on the inside; this was the diptych (see also Diptychs, Triptychs, Polyptychs, Plaques, p. 168).

Our interest here is in the fact that later models were sometimes made of ivory. The less expensive ones had a plain outside ivory surface, whereas the more costly ones were sculpted with domestic scenes, mythological figures, historical events, or in later times, scenes representing Christian themes.

By A.D. 300, the inauguration of Roman consuls had become established as an occasion for ostentation and lavish expenditure. One of the favored gifts which a new consul sometimes presented to each important official, especially the emperor who had appointed him, was an ivory writing tablet diptych. The two outer surfaces were finely carved by the best sculptors, portraying whatever scenes and historical inscriptions the consul desired, usually concerning the inaugural event itself. This practice continued until A.D. 541, the year of the last consul, Basilius, inaugurated at Constantinople. Almost all of the known surviving consular diptychs ended up in Europe's museums—in Paris, Milan, Berlin, Liverpool, London, and more than a dozen other cities. A few are still in private collections.[22]

Fine non-official writing tablets (private diptychs) were once given by affluent families to celebrate a joyous personal event, such as a marriage or an unexpected recovery from a serious illness. As with many consular diptychs, the two hinged sections were often separated and they found their way not only into different collections, but even into different countries. Although the ivory diptych began as a pair of waxed writing tablets, the greater appeal as an art form led inevitably to a decline in their use as writing surfaces.

Before leaving this topic, it might be well to note that the inside waxed surface of diptychs was not the sole provision for writing on ivory. In addition to the ancient Chinese memorandum tablets in the form of an ivory slab (see Memorandum Tablets and Pads, p. 189), a small number of diptychs were made with one or more ivory pages between the two waxed tablets.[23] On these ivory sheets, either permanent writing in ink could be registered or erasable notations could be entered in pencil (see also Books of Ivory, p. 90).

23 ✤ MISCELLANEOUS

The full variety of utilitarian products incorporating ivory will probably never be known. There are, in particular, so many different kinds of products with ivory handles that it seems unlikely we will ever discover them all. Some kinds made in small quantities may have disappeared entirely, perhaps thrown onto trash heaps of the past.

In addition to the ivory articles treated in the twenty-two preceding categories, other known ivory articles will now be considered in this miscellaneous final grouping. A number of tools for changing the form or appearance of material, which were not relevant to the preceding categories, are grouped together in a single section.

Admission Tickets During the Roman Empire, admission tickets to the theater were sometimes in the form of a small flat piece of ivory called a tessera. The Joseph Mayer Museum in Liverpool, England, has a circular ivory admission ticket for seat number 8 of a Roman theater.[1]

Batons Ivory ceremonial batons for Chinese dignitaries date back to the Han dynasty (206 B.C. – A.D. 221).[2] A celebrational baton with ivory end-sections and colored-glass inserts is shown in Figure 61; the inscription in German is translated as "Christian Social Workers Association of Bensheim. Made for the 25-Year Founders' Celebration of 30 May 1897." (See also Conductors' Batons, p. 137.)

Boat Fittings Alaskan Eskimos made several kinds of boat fittings from walrus ivory. Figure 62(c) shows a boat spur; this writer has been informed that a number of these were attached horizontally to the side edges of a boat in northern waters to aid the passage of the vessel through ice sheets.

Chariot Ornamentation Emperors of the Chou dynasty in China (eleventh to third century B.C.) rode in chariots that were sometimes decorated with ivory inlays.[3] It has been said that the chariot door was sometimes crafted entirely from joined sections of ivory. Readers may find it of interest to note that German artisans in Stettin crafted an entire miniature coach from ivory; they had it pulled by "trained" fleas, so providing an entertaining curiosity.[4]

Corkscrews With the increasing use of corks as bottle stoppers in the seventeenth century, corkscrews naturally became an important accessory. Although corks were primarily used for wine bottles, they were also occasionally used in bottles for perfume, liquid medicines, and so on. Handles made of ivory and other materials were crafted in a great variety of shapes and representational forms, especially human and animal figures. Figure 63 includes a small ivory-handled corkscrew; it was probably for the corks in small bottles of perfume. Collectors browsing through antique shops may still be able to obtain corkscrews with a plain ivory handle.

De-knotters Some ancient carvers took knots seriously enough to provide a special form of pointed ivory instrument for loosening a knot. An early Chinese de-knotter, richly carved and embellished, is at the Metropolitan Museum of Art in New York. Of solid ivory, it is about 3 inches (7.5 cm.) long and has a flat blade that tapers down to a point; the point is gradually inserted into the knot until the knot eventually becomes loosened. This particular example, however, was intended to be more symbolic than useful. Worn suspended from the waist belt of a young man, it symbolized that he was entering manhood and could now, so to speak, unravel life's problems.[5]

Fertilizer Wilson and Ayerst reported that ivory waste, such as shavings and sawdust, was sometimes used as fertilizer (no specific countries were named).[6] In a personal communication to the present author, Ayerst said that he knew an Indian ivory merchant in Mombasa who used ivory dust as fertilizer for his rose plants; he believed that his roses had become the best in the area.[7]

Fids Looking like an oversized bodkin, a fid was a large pin used by seamen for separating the twisted strands at the end of a rope prior to splicing them to the separated strands of another rope. The carved ones were usually made of whalebone. The earliest known dated fid shows the scrimshander's name and the year 1774.[8] Because of their large size and minor function, they were usually carved from whalebone, but some may have been made of walrus ivory or whale-tooth ivory. Whalebone fids can be seen at the Mystic Seaport Museum in Mystic, Connecticut.

Gavels A gavel was usually made of hard wood, though some were crafted from ivory, especially when intended to be used for such prestigious occasions as fine art auctions. Some gavels have consisted of only the rapper, without a handle. Figure 61 shows an ivory gavel crafted long ago for or concerning "BPOE 1378." Assuming that this referred to the Benevolent and Protective Order of Elks, Chapter 1378, this writer was curious to know the fateful event that brought such a fine gavel to the curio shop in which it was displayed. Had it been lost or stolen, or sold when equipment from a discontinued chapter was dispersed? A still-existing Chapter 1378 at Redondo Beach, California, was traced, but not even the earliest members could recall having ever seen an ivory gavel there.

Handles for Bags, Boxes, Buckets, Sleds Alaskan Eskimos once carved walrus-ivory handles for holding bags, boxes, and buckets, each kind of handle slightly differing in form. Figure 62(b) shows a box handle, simply but decoratively incised. A curved handle for a bucket is shown in Figure 62(a). Eskimos made ivory and bone drag-handles for pulling a light carrier on ice.[9]

Howdahs In nineteenth-century India, royalty riding on an elephant's back sat on a howdah that was sometimes made of ivory.[10] Such a spectacular seat was composed of large jointed sections of ivory blocks.

Jeweler's Polish In addition to various other uses of ivory waste, ivory sawdust has been used as a polishing agent. This function was still being reported in northwest India at the end of the nineteenth century.[11]

Ladies' Portable Kits Small kits with useful articles made of ivory were favorite gifts for ladies during the nineteenth century. They were carried in a purse and might contain from three to six tiny articles. Figure 65 shows such a kit, containing a refillable pencil, bodkin, thimble, and needlecase.

Latch Pegs Small latch pegs made of wood, bone, or ivory were sometimes used in Japan for keeping closed the hinged lid or front opening of a box. They were also used for keeping closed a folding cardboard protective case for a book. Figure 65 includes an ivory latch-peg used in Japan for just such a cardboard case; the peg, suspended by a slender silk ribbon attached to one side of the case, was inserted down into a silk loop on the side that opened for removing or replacing the book.

Medals Medals of honor awarded in ancient times were sometimes made of ivory. They date back to the first century A.D. according to a reference to them in the writings of Diodorus Siculus, a Greek historian who was a contemporary of Julius Caesar and Augustus.[12]

Miniature Replicas One of the miscellaneous uses of ivory was for furnishing tiny replicas of articles that had symbolic significance for certain crafts, trades, or fraternal organizations. For example, members of the Masonic fraternal society could obtain a set of miniature ivory symbolic replicas of masons' tools, for display on a desk or for framed display on a wall. Figure 64 shows a set and case made in Japan.

Money Figure 65 shows an example of an ivory coin. This 25-cent piece was one of several denominations of coins from the Cocos Islands (also known as the Keeling Islands), in the Indian Ocean. Made in England, these coins were apparently authorized in 1910 and then were the currency used from 1913 until about 1956, when control of the islands' administration passed from Singapore (a British crown colony) to Australia. One side (shown) is inscribed "Keeling Cocos Islands 1910"; the other side shows the coin's serial number, the denomination, the year 1913, and the name of the early settler-owner, J. S. Clunies Ross. The denominations were 5 cents, 10 cents, 25 cents, 50 cents, 1 rupee (100 cents), 2 rupees, and 5 rupees; only a few thousand coins of each value were issued.

Oliphants An oliphant (early English name for an elephant) was the pointed section cut from a tusk with the solid inner substance removed. This conical-shaped section was cut about 6 to 12 inches (15–30 cm.) long.[13]

If the point was also cut off, the hollowed oliphant, now open at both ends, could be used as a sounding horn (as a signal during a hunting chase, to sound an alarm inside a castle, or to announce the beginning of church services). With a cap placed tightly over both ends, the oliphant served as a gunpowder horn.

If the tip was not cut off, the horn was used as a container. Many were drinking horns and often were finely sculpted on the entire outside surface. Ecclesiastical oliphants contained small relics and were frequently seen in Western Europe and England.

Probably the most unusual function of oliphants was their use in England as tenure horns; these were inscribed as evidence that land or an appointment to an official position had been granted. The recipient's tenure horn was later treasured as a family heirloom and handed down through successive generations. Examples can be seen at the Victoria and Albert Museum in London; one presented by Henry I in the twelfth century, when he granted the church certain lands, is in Carlisle Cathedral (northwest England).

Palanquins In ancient China, one of the drains upon the supply of elephant ivory resulted from the use of whole tusks to make ivory palanquins. These were single-seat litters supported between two parallel horizontal poles which protruded at the front and back; the burden was carried on the shoulders of litter bearers. High military and civil officials honored an appointment with the emperor by being carried into the royal chamber while seated on their impressive ivory palanquin.[14]

India produced several kinds of palanquins, some of them resembling a couch or bed. Thin ivory plates covered a wooden frame and were richly engraved with designs.[15]

Pickwicks A long, slender metal pin, sometimes made with an ivory handle, was used for picking up the shortened wick of an oil lamp. Some of them were made by American seamen from whale tooth. Several can be seen at the Mystic Seaport Museum.

Pins The earliest pins used for holding garments together are thought to have been thorns and sharp slender fishbones. Although straight pins were occasionally made of ivory, and were therefore rather attractive, ivory pins could not be as serviceable as the ones made of metal. The earliest ones were found in ancient Egyptian tombs, dating before 3100 B.C. Besides being employed as garment fasteners, modern ones have had various uses.

Pocketknives and Belt Knives Besides the ivory handles of knives used as eating utensils or as daggers, general-use portable knives, such as pocketknives and belt knives, sometimes had ivory handles. Figure 63 shows an ivory-handled pocketknife carved as a human figure; it was bought in an antique shop in Greece.

Riding Whips The flexibility of thin strips of fresh ivory is demonstrated by the fact that long thin strands have been used as riding whips (see also Whips, p. 187).[16] The riding crop—a short whip in the form of a baton-like stock (usually with a short lash at the end)—was sometimes cut from ivory.

Rope Twisters Whereas a fid served to open the intertwined strands at the end of two ropes so that the ends could be sliced together (see Fids, p. 196), a rope twister was used when manually twisting several long strands to form a rope. They were crafted from whalebone.[17]

Rulers In China, ivory measuring-sticks date back to the Tang dynasty (A.D. 618–906). Ivory rulers made later in Shanghai can be seen in the Chinese Exhibit at Chicago's Field Museum of Natural History. These have 10 units, each about 1 inch (2.5 cm.) long and further divided into tenths. Interestingly, but not unexpectedly, artistic impulse was expressed even in something as naturally plain as a measuring stick; the backs of Chinese measuring instruments were sometimes etched with delicate floral designs and birds in color. By the nineteenth century, ivory rulers were being furnished by India and several countries of the Western world. Figure 63 shows an old 12-inch (30 cm.) folding ivory ruler (note the reverse numbering).

Scepters Historical accounts tell that after the downfall of Tarquin (tyrant and last king of Rome, ruling 534–510 B.C.), his ivory scepter was given to his friend and supporter, Lars Porsena, the legendary Etruscan. The Roman Senate presented an ivory scepter to Eumenes II, King of Western Asia Minor in the second century B.C. Centuries later, the Vikings carved scepters from the long slender tusks of narwhals.[18] Several finely carved Chinese scepters from the eighteenth century are at the Guimet Museum in Paris.[19]

Somewhat similar to the regal or imperial scepter is the mace, now merely a ceremonial staff, but once a clublike weapon of war. An ivory ceremonial mace was elaborately carved in London by the Carrington Company for the Speaker of the Kenyan Legislative Assembly.[20]

Size This writer has seen only a single reference indicating that one use of ivory sawdust was for the production of size, a gelatinous preparation used for filling the porous surface on paper and cloth.[21]

Sled Runners The Pitt Rivers Museum of the University of Oxford in England has a pair of Eskimo sled runners that were made by splitting a walrus tusk in half longitudinally.[22] Figure 61 shows a bone section (broken) of an Alaskan sled runner. Some believe that the Vikings used animal shinbones as ice skates.

Snuff Graters Similar to the use of graters for reducing tobacco leaves to small bits suitable for pipes, the more finely pulverized form for sniffing gave rise to a demand for ivory snuff graters during the seventeenth and eighteenth centuries. Before the French Revolution, Dieppe was a prominent center for the carving of ivory snuff graters (and other ivory products). Many were made in England, Germany, Holland, and Belgium. Richly carved snuff graters, mostly with human figures, can be seen at the Victoria and Albert Museum and the Wallace Collection in London.[23]

Snuff Knives In a treatise on mariners' scrimshaw crafts of the past, there is an illustration of two small folding snuff knives made of ivory or bone.[24] A written inquiry to the author of that work regarding the function of the small knives brought a reply that such a conveniently sized knife might have been used to break up compacted snuff inside the container, or perhaps was a snuff stick used to apply small portions of snuff to the gums as a stimulant.

Spatulas Ivory spatulas were found in Egyptian predynastic tombs (before 3100 B.C.). The section on combs (p. 161) mentioned that ancient cremation sites in southern Spain yielded ivory artifacts of Phoenician style, possibly dating as early as 700 B.C.; one of these ivories was a spatula. The functional end is oval shaped; the handle is in the form of a swan's head and neck. A similar spatula was found in a Punic grave at Carthage in North Africa.[25] The function of such ivory burial spatulas is unknown.

Toggles In the category on artistic ivories, it was shown that the Japanese netsuke was originally a plain toggle used for suspending certain small useful articles from a man's waist belt (see p. 81). Toggles have had other uses, of course, and we note that Alaskan Eskimos carved walrus-ivory toggles in order to "keep carrying-straps and boat-lashings from slipping."[26]

Tools Because the word "tool" has such a broad range of applications, a separate category has not been provided among the twenty-three groupings of ivory products; instead, most tools have already been treated under the specific category to which they were relevant. Any tools not relevant to those categories have been assigned to this general grouping of miscellaneous ivory products. The ones intended to be used solely, or mainly, for changing the form or appearance of material are considered below. Other kinds of tools are listed in separate sections of the miscellaneous category. (Note that in what follows, the designation "Eskimo" does not imply that such tools were made only by Eskimos.)

ADZ: This is a cutting tool, usually with a thin arched blade set at right angles to the handle. An Eskimo adz head of walrus ivory (stained black) is shown in Figure 62(d).

AWL: An Eskimo awl of walrus ivory is shown in Figure 62(e). An awl is used for marking surfaces or piercing small holes; the tool shown was probably adequate for soft materials only. The two holes suggest that this awl, like some bodkins, may also have been used as a threader for passing an attached cord through a slender channel in clothing.

BOW DRILL: Alaskan Indians and Eskimos made bow drills for which the hand-operated bow was sometimes made of bone or ivory.[27]

GOUGER CHISEL: A gouger chisel with an ivory handle (origin uncertain) is shown in Figure 62(f).

HAMMER: Eskimos made entire hammers from walrus ivory.

HATCHET: Reportedly, Eskimos crafted these from solid walrus ivory; it would seem that the cutting edge could be used only on softer materials.

ICE PICK: A solid walrus-ivory ice pick, made by Eskimos, is shown in Figure 62(g).

[202] Part Two : Uses of Ivory

MALLET: Nineteenth-century American whaler-seamen made whalebone mallets for softening or flattening moderately hard material.[28] One that was used for tenderizing meat is at the Mystic Seaport Museum.

SCISSORS: Antique dealers have mentioned that occasionally ladies' small scissors had handles cut from ivory or bone. This writer has not seen one, nor even a photograph of one.

SNOW-SHOVEL EDGE: Eskimos sometimes made the bottom edge of a shovel from walrus ivory.[29]

WEDGE: It seems surprising to learn that even a sharp ivory edge could be hard enough to split cracked logs, as has been reported in some writings from Africa.

Whisk Brooms and Dusters Whisk brooms and furniture dusters have been made with an ivory or whalebone handle. Figure 63 shows a broom—perhaps nineteenth century—in which the straw fibers are very firm, as in whisk brooms. But several features of this brush are not typical of a whisk broom: (1) the circular brush; (2) the time-consuming cut-through carving work; (3) other artistic embellishment, including gilt paint, and glass insets near the top; (4) the pick (also made of ivory) that is held in a vertical slot at the top of the ivory handle. The work appears to be Oriental, perhaps Indian or Southeast Asian. It is a curious article indeed.

Nineteenth-century American whaler-seamen sometimes carved a whalebone handle for a furniture duster.[30] One can be seen at the Mystic Seaport Museum.

Whistles Small toy whistles of ivory are treated earlier (see p. 182). Large ivory whistles, used as signaling devices, were sometimes carved quite finely. An example can be seen at the Santa Barbara Historical Society Museum in California.

In a few of the twenty-three usage categories, it has already been indicated that the exact function of some of the products shown remains a little uncertain. Uncertainty usually increases, of course, when only a part of the original product is available. A few examples of even greater uncertainty are presented now; for most of these, even the correct category seems doubtful. They are shown in Figure 66, labeled (a) to (e), and are described below accordingly.

(a) This attractive handle, with the functional end piece missing, seems to defy identification. The ornamentation precludes holding it as a pen. The seller in Italy thought it had been some kind of sewing tool.

(b) Could this be a finial? For added attractiveness, ivory was used for various fittings, such as small detachable pieces on household furnishings and appliances. The figure shows a machine-turned cylinder, one piece of which is an attractive screw for the matching section. These may once have been fitted at the top of a lamp for securing the small center ring of a lampshade. Plain ivory screws, too, were made in earlier times, usually for use on the exterior of ivory boxes.

(c) While this could also be a finial, it is more likely a cabinet knob.

(d) The figure shows a set of ivory rings which may once have secured the lenses at the front and back ends of a telescope.

(e) Here are two similar examples (of ivory or bone) of a device with a long string wound around and attached to a small H-shaped slat. The other end of the string is attached to a peg; a second peg can be moved along the string. Because the peg and slat are attached to the string, it seems unlikely that such a device is a sewing instrument. But it could be used to measure off equal lengths of cloth from a roll. Having separated the two pegs to the desired length, one could then mark that length for the number of pieces needed. The allowable separation distance is quite ample: almost 6 feet (1.8 m.) on one device and almost 15 feet (4.6 m.) on the other (the one that advertises whisky).

APPENDIX: MUSEUM COLLECTIONS

These listings represent most of the known collections of ivory art and ivory utilitarian products. However, such collections are sometimes reduced in size, or even entirely transferred to other museums. Furthermore, any museum's ivories may have been stored away, entirely or in part, rather than being on exhibit at the time you visit the museum. If you have a card from your hometown's museum showing Contributing Membership status, it will sometimes induce a museum official to show you part or all of a stored group of ivories.

Foreign names are translated into English here. In a few instances, where literal translation alone might be unclear, both the foreign name and the English equivalent are provided. If a museum has two different names, both are given.

For readers who may need to recognize foreign references to "elephant" and "ivory," and for collectors visiting antique shops and museums in foreign countries, here is a selective list of terms:

French—Elephant/Ivoire
German—Elefant/Elfenbein
Italian—Elefante/Avorio
Russian—Slon/Slonovaya kost
Spanish—Elefante/Marfil

AFGHANISTAN
KABUL: National Museum of Afghanistan

ARGENTINA
ROSARIO: City Museum of Decorative Art

AUSTRIA
VIENNA: Austrian Museum of Fine Arts
Museum of the History of Art

BELGIUM
BRUSSELS: Museum of Decorative and Industrial Arts
Royal Museum of Central Africa
LIEGE: Curtius Museum
Diocesan Museum
MAASEIK: Stedelijk Museum

BRAZIL
RIO DE JANEIRO: National Historical Museum

CANADA
MONTREAL: Montreal Museum of Fine Arts

REPUBLIC OF THE CONGO
BRAZZAVILLE: National Museum of the Congo

DENMARK
COPENHAGEN: Museum of Decorative Arts
National Museum

EGYPT
CAIRO: Coptic Museum
Egyptian National Museum

ENGLAND
BIRMINGHAM: City Museum and Art Gallery
LIVERPOOL: Free Public Museum
Joseph Mayer Collection
LONDON: British Museum
Jewish Museum
Victoria and Albert Museum (formerly Kensington Museum)
Wallace Collection, Hertford House
LUTON: Wernher Collection
MANCHESTER: John Ryland Library
OXFORD: Ashmolean Museum
Bodleian Library
SALISBURY: Salisbury and South Wiltshire Museum
STOKE-ON-TRENT: City Museum and Art Gallery

FRANCE
AMIENS: Collection Charles de L' Escalopier, City Library
ANNECY: Montrottier Palace
BEAUNE: Hotel-Dieu Museum
COMMERCY: City Museum
DIEPPE: Old Chateau Museum
DIJON: Museum of Fine Arts
METZ: Metz Museum
PARIS: Cluny Museum
Ennery Museum
Guimet Museum
Louvre Museum
Museum of Decorative Arts
National Library, Manuscripts Department and Medals Collection
SAINT GERMAIN: City Museum
National Museum of Antiquities

GERMANY*
AACHEN: Suermondt Museum
BABENHAUSEN: Fugger Museum
BERLIN: Arts and Crafts Museum (Kunstgewerbe Museum)
Museum of Islamic Art
Sculpture Department Museum
BRUNSWICK: Herzog Anton Ulrich Museum
History and Ethnology Museum
COLOGNE: Schnütgen Museum, St. Cecilia Church
DARMSTADT: Hesse Public Museum
DRESDEN: Green Dome (Grünes Gewölbe)
ERBACH: German Ivory Museum
HAMBURG: Museum of Arts and Crafts
HARBURG: Library and Art Collection
MICHELSTADT: Ivory Museum
MUNICH: Art Collection of the State Library
Bavarian National Museum
Treasury of the Munich Residence
NUREMBERG: German National Museum
TRIER: Cathedral Treasury Room (Domschatzkammer)
WALLDÜRN: Walldürn Ivory Exhibit

GREECE
ATHENS: Agora Museum
National Archeological Museum
National Historical Museum
HERAKLION, CRETE: Archeological Museum
SAMOS: Archeological Museum
THEBES (THIVAI): Archeological Museum

HUNGARY
BUDAPEST: Iparmüvészeti Museum

INDIA
BOMBAY: Prince of Wales Museum of Western India

*No distinction is made here between the two Germanies; almost all of the significant ivory collections are in West Germany.

Delhi: Crafts Museum, Thapar House
Hyderabad: Salar Jung Museum
Trivandrum: Art Museum

IRAN
Teheran: National Art Museum

IRAQ
Baghdad: National Museum

ITALY
Agrigento, sicily: Diocesan Museum
Arezzo: Government Museum of Medieval and Modern Art
Avellino: Irpino Museum
Bologna: City Museum
  Town Palace, Communal Art Collection
Brescia: City Museum
Florence: Bargello National Museum
  Pitti Palace, Silver Museum
Grosseto: Diocesan Museum of Sacred Art
Milan: Archeological Museum
  Cathedral Treasury Museum
  Sforzesco Castle Museum
Monza: Basilica Treasury
Naples: Museum of the Duke of Martina
  National Archeological Museum
  National Museum and Gallery, Capodimonte Palace
Palermo, sicily: Palace of the Normans
Ravenna: Cathedral Museum
  City Museum
  National Museum
Rome: African Museum
  Barberini Palace Library
  Church Museum of the Roman College
  Vatican Museums
Salerno: Cathedral Museum
Turin: City Museum of Ancient Art

JAPAN
Kyoto: National Museum
Nagano: Kitano Art Museum
Nara: Horyuji Temple
  Shosoin Imperial Treasury Museum
Tokyo: National Museum of Tokyo

MEXICO
Puebla: José Luis Bello and Gonzalez Museum

SAN MARINO
San marino: City Museum

SCOTLAND
Glasgow: Glasgow Art Gallery and Museum

SOUTH AFRICA
Durban: Durban Museum and Art Gallery

SPAIN
Burgos: Cathedral Museum
  Provincial Archeological Museum
Cordova: Provincial Archeological Museum
Madrid: Lazaro Galdiano Museum
  National Archeological Museum
  National Museum of Painting and Sculpture (The Prado)
Toledo: Museum of the Duke of Lerma Foundation
  Valencia Institute of Don Juan
Zaragoza: Diocesan Museum

SWEDEN
Balsta: Skoklosters Castle

SWITZERLAND
Basel: Historical Museum
Zurich: Swiss National Museum

TAIWAN (Republic of China)
Taipei: National Museum of History
  National Palace Museum

UNITED STATES
Anchorage, alas.: Anchorage Historical and Fine Arts Museum
Baltimore, md.: Walters Art Gallery (European and Egyptian; occasional display: Japanese)
Boston, mass.: Museum of Fine Arts
Chicago, ill.: Field Museum of Natural History (African, Chinese, Japanese)
  Oriental Institute Museum (Ancient Palestinian)
Fairbanks, alas.: University of Alaska

Museum (Eskimo; see also Anchorage museum listed above)
FLINT, MICH.: Flint Institute of Arts (Chinese)
GLENDALE, CALIF.: Forest Lawn Museum
GREENWICH, CONN.: Bruce Museum (occasional display only: Japanese, Chinese, Indian)
KENOSHA, WIS.: Kenosha Public Museum (Japanese)
LOS ANGELES, CALIF.: Fowler Museum of Decorative Arts (Chinese, European)
Los Angeles County Museum of Natural History (American scrimshaw, African horn, and African long trumpet)
MIAMI, FLA.: Vizcaya Dade County Art Museum (occasional display only: small Oriental collection)
MILWAUKEE, WIS.: Milwaukee Public Museum (Japanese, Chinese, European)
MYSTIC, CONN.: Mystic Seaport Museum (American scrimshaw)
NANTUCKET, MASS.: Nantucket Historical Association (American scrimshaw)
NEW BEDFORD, MASS.: Whaling Museum (American scrimshaw)

NEW YORK CITY: American Museum of Natural History (Japanese)
Metropolitan Museum of Art
NEWARK, N. J.: Newark Museum (occasional display only: Japanese, Chinese, Indian, African, and some European)
NEWPORT NEWS, VA.: Mariner's Museum (American scrimshaw)
SALEM, MASS.: Peabody Museum (American scrimshaw)
SAN FRANCISCO, CALIF.: Maritime Museum (American scrimshaw)
SANTA CLARA, CALIF.: de Saisset Art Gallery and Museum (occasional display only)
SEATTLE, WASH.: Museum of History and Industry (Alaskan)
Seattle Art Museum (occasional display only: Japanese, Chinese, African)
SNYDER, TEX.: Diamond M Foundation Museum (Chinese and Japanese)
SPRINGFIELD, MASS.: George Walter Vincent Smith Art Museum (Japanese)
WASHINGTON, D.C.: National Museum of Natural History, Department of Anthropology (for research study only: Eskimo, African, Chinese, Japanese)

NOTES

PART ONE

CHAPTER I

1. Sergio Bosticco, "Ivory and Bone Carving," *Encyclopedia of World Art*, vol. 8, New York: McGraw-Hill, 1963, cols. 763–64 and Pl. 236.
2. George C. Williamson, *The Book of Ivory*, London: Frederick Muller, 1938, pp. 118–19, 124.
3. Gordon Loud, *The Megiddo Ivories*, Chicago: University of Chicago Press, 1939, Pls. 1–63.
4. R. D. Barnett, *A Catalogue of the Nimrud Ivories, with Other Examples of Ancient Near Eastern Ivories*, London: British Museum, 1957, pp. 123–24.
5. *Ibid.*, p. 16.
6. Taiji Maeda, "Ivory and Bone Carving," *Encyclopedia of World Art*, vol. 8, New York: McGraw-Hill, 1963, cols. 779–84.
7. Berthold Laufer, *Ivory in China*, Chicago: Field Museum of Natural History, 1925, pp. 7–9, 52–53.
8. Barry C. Eastham, *Chinese Art Ivory*, Ann Arbor, Mich.; Ars Ceramica, 1976, p. 61.
9. Laufer, pp. 68–69 and Pl. VIII.
10. George F. Kunz, *Ivory and the Elephant: In Art, in Archeology, and in Science*, New York: Doubleday, 1916, pp. 266–78.
11. Ananda K. Coomaraswamy, *The Arts and Crafts of India*, New York: Farrar Straus, 1964, chap. 7.
12. Edgar Thurston, *Monograph on the Ivory Carving Industry of Southern India*, Madras: Government Press, 1901.
13. E. B. Havell, "Ivory Carving in Madras," *Journal of Indian Art*, vol. 2 (1888), pp. 20–21.
14. T. P. Ellis, "Ivory Carving in the Punjab," *Journal of Indian Art*, vol. 9 (1901), no. 75, pp. 45–52.
15. M. Digby Wyatt, *Industrial Arts of the Nineteenth Century at the Great Exhibition, 1851*, London: Day & Son, 1853, vol. 1, Pl. 21 text.
16. Kunz, pp. 400–401.
17. Williamson, pp. 73–74.

18. William Maskell, *Ivories Ancient and Medieval,* London: Chapman & Hall, 1870, pp. 44-45.

19. Carson Ritchie, *Modern Ivory Carving,* New Jersey: A. S. Barnes, 1972, pp. 98-109; p. 108 shows a spectacular dressing table mirror-frame and base.

20. Massimo Carrà, *Ivories of the West,* New York: Hamlyn, 1970, pp. 82-84, 100-101.

21. Kunz, pp. 508-9.

22. Ritchie, pp. 68-78, 110-24.

23. Charles Holtzapffel, *Turning and Mechanical Manipulation,* 6 vols., 2nd ed., London: Holtzapffel & Co., 1846.

24. Ritchie, pp. 125-27; sections from the Jenkins frieze are shown on pp. 139, 142.

25. Norbert Beihoff, *Ivory Sculpture through the Ages,* Milwaukee: Public Museum, 1961, p. 24.

26. Derek Wilson and Peter Ayerst, *White Gold: The Story of African Ivory,* New York: Taplinger, 1976.

27. Ernest D. Moore, *Ivory, Scourge of Africa,* New York: Harper, 1931.

28. Heinrich Brode, *Tippo Tib: The Story of His Career in Central Africa,* London: Edward Arnold, 1907.

29. William B. Fagg, *Afro-Portuguese Ivories,* London: Batchworth Press, 1959, Pl. 28.

30. Dorothy Jean Ray, *Eskimo Art: Tradition and Innovation in North Alaska,* Seattle: University of Washington Press, 1977, pp. 38-39, 68.

31. Dorothy Jean Ray, *Artists of the Tundra and the Sea,* Seattle: University of Washington Press, 1961, pp. 26-35.

32. Ritchie, pp. 35-36, 60-65.

33. Kunz, pp. 96-97.

CHAPTER II

1. Schuyler Cammann, "Ivory and Bone Carving," *Encyclopedia of World Art,* vol. 8, New York: McGraw-Hill, cols. 757-63.

2. Otto Pelka, *Elfenbein,* Berlin: Schmidt & Co., 1920, p. 10.

3. Robert Webster, "Ivory, Bone, and Horn," *The Gemmologist,* vol. 27 (1958), no. 322, pp. 91-98.

4. Charles P. Woodhouse, *Ivories: A History and Guide,* New York: Van Nostrand Reinhold, 1976, p. 19.

5. Schuyler Cammann, "Carvings in Walrus Ivory," *Pennsylvania University Museum Bulletin,* vol. 18 (1954), no. 3, pp. 2-31.

6. Laufer, pp. 40, 49-50.

7. Cammann, "Carvings in Walrus Ivory"; an abstinence plaque is shown facing p. 22.

8. Thomas K. Penniman, *Pictures of Ivory and Other Animal Teeth, Bone, and Antler; with a Brief Commentary on Their Use in Identification,* Oxford: Oxford University Press, 1952, p. 29.

9. Kunz, pp. 292-98.

CHAPTER III

1. E. Norman Flayderman, *Scrimshaw and Scrimshanders; Whales and Whalemen,* New Milford, Conn.: N. Flayderman & Co., 1972; examples are shown on pp. 8-9, 40-41.

2. *Ibid.,* pp. 18-24.

3. Lilla Perry, *Chinese Snuff Bottles,* Tokyo and Rutland, Vt.: Charles E. Tuttle, 1960, p. 125.

4. Schuyler Cammann, "Account of Hornbill Ivory," *Pennsylvania University Museum Bulletin*, vol. 15 (1950), no. 4, pp. 19–47.

5. Schuyler Cammann, "Jewelry from a Bird," *Antiques*, vol. 66 (1954), July, pp. 34–36.

6. Edward Albes, "Tagua—Vegetable Ivory," *Bulletin of the Pan American Union*, vol. 37 (1913), pp. 192–208.

7. Acosta Solis, *La Tagua*, Quito, Ecuador: Publicaciones Cientificas, 1944.

8. "Doum Nuts of Commerce," *Scientific American*, vol. 116 (1917), p. 129.

9. Alfred Maskell, *Ivories*, London: Methuen, 1905; rpt. ed., Tokyo and Rutland, Vt.: Charles E. Tuttle, 1966, p. 480.

10. Much of the information presented here about synthetic substitutes is based on articles by Virginia Botsford in "Celluloid," *Collectibles Monthly*, May, June, July, 1978.

11. "Artificial Ivory," *Scientific American*, vol. 64 (1891), p. 73.

CHAPTER IV

1. L. M. Stubbs, "Ivory Carving in the Northwest Provinces and Oudh," *Journal of Indian Art*, vol. 9 (1901), no. 75, pp. 41–45.

2. Charles Holtzapffel, *"Materials,"* p. 145, *Turning and Mechanical Manipulation*, vol. 1, 2nd ed., London: Holtzapffel & Co., 1846.

3. O. M. Dalton, *Catalogue of the Ivory Carvings of the Christian Era with Examples of Mohammedan Art and Carvings in Bone, in the British Museum*, London: British Museum, 1909, Pl. CXIV.

4. Alfred Maskell, p. 479.

5. Kunz, p. 263.

6. Williamson, p. 12.

7. Penniman, p. 15.

8. Holtzapffel, vol. I, pp. 153–54.

9. *Ibid*.

10. Julius Pratt, *Centennial of Meriden*, June 10–16, 1909, pp. 327–28. The exhibited ivory sheet was reported in *Great Exhibition of the Works of Industry of all Nations, 1851. Official Descriptive and Illustrated Catalogue*, London: Spicer Brothers, 1851. In vol. 3, United States Section, p. 1468, item 567 reads: "Pratt, J. and Company, Meriden, Connecticut. Inventor. Specimen of ivory veneer, cut by machinery. The machine by which these veneers were sawn is a recent invention of the exhibitors. One piece of this veneering is 12 inches wide and 40 feet long, sawn from a single tusk." The discrepancy between the exhibitor's report (14 inches by 52 feet) and the catalog listing is puzzling.

11. Alfred Maskell, p. 478.

12. Margaret Longhurst, *English Ivories*, London: Putnam's Sons, 1926, p. 61. For a detailed description of early reproduction machines, see G. B. Hughes, "Mechanical Carving Machines," *Country Life*, Sept. 23, 1954, pp. 980–81.

13. Carson Ritchie, *Ivory Carving*, London: Arthur Barker, 1969.

CHAPTER V

1. N. S. Baer, and others, "The Effect of High Temperature on Ivory," *Studies in Conservation*, vol. 16 (1971), no. 1, pp. 1–8.

2. Eugen Philippowich, *Elfenbein*, Braunschweig, Germany: Klinkhardt & Biermann, 1961, p. 325.

3. M. H. Spielmann, "Art Forgeries and Counterfeits: Ivories," *Magazine of Art*, vol. 27 (1903), pp. 549–50.

[212] *Notes*

4. The first three examples are from Williamson, pp. 210, 212.
5. Carrà, p. 70 and Pl. 32.
6. *Ibid.*, p. 78 and Pl. 36.
7. William Maskell, pp. 104–7.
8. Raymond Bushell, *The Netsuke Handbook of Ueda Reikichi*, Tokyo and Rutland, Vt.: Charles E. Tuttle, 1961, p. 117.

CHAPTER VI

1. South Kensington Museum, *A Description of the Ivories Ancient and Medieval in the South Kensington Museum, with a Preface by William Maskell*, London: Chapman & Hall, 1872, preface.
2. Ritchie, *Modern Ivory Carving*, pp. 161–63.
3. Nancy Armstrong, *The Book of Fans*, New York: Mayflower Books, 1978, p. 84.
4. Alfred Maskell, p. 492.
5. Kunz, p. 262.
6. John F. Mills, *The Care of Antiques*, New York: Hastings House, 1964, pp. 42–44.
7. Clayton Holiday, "The Conservation of Ivory and Bone," *Bulletin of the South African Museums Association*, vol. 9 (1971), no. 10, pp. 561–74.

CHAPTER VII

1. Williamson, pp. 206–9.
2. Ritchie, *Ivory Carving*, pp. 128–30.
3. Robert Webster, "Vegetable Ivory and Tortoise Shell," *The Gemmologist*, vol. 27 (1958), no. 323, pp. 103–7.
4. George Savage, *The Antique Collector's Handbook*, London: Barrie & Rockliff, 1959, pp. 229–33.

PART TWO

1 ARTISTIC IVORIES

1. Examples can be seen in the following works: Beihoff, pp. 61, 63; Alfred Maskell, Pls. 30–34; O. Beigbeder, *Ivory*, New York: Putnam's Sons, 1965, Figs. 22, 29.
2. Shown in Williamson, facing p. 212.
3. *Ibid.*, pp. 188–89.
4. Many of these are shown in Dalton, Pls. CI, CII, CIII; other portrait medallions are shown in Therle Hughes, *Small Antiques for the Collector*, New York: Macmillan Co., 1964, Fig. 32.
5. Beautiful examples are shown in Warren E. Cox, *Chinese Ivory Sculpture*, New York: Crown, 1946, Pl. 20.
6. Eastham, p. 39.
7. Elaborate examples are shown in Eastham, Pls. 12, 13, 38, 39.
8. Fagg, Pl. 28.
9. Shown in Tardy and others, *Les Ivoires*, Paris: Tardy, 1966, p. 244.
10. Dalton, Pl. CV.
11. A seated ivory skeleton $1\frac{13}{16}$ inches (45 mm.) high is shown in Bushell, Fig. 64.
12. Shown in Beihoff, p. 76.
13. Beigbeder, Figs. 113–16; Dalton, Pl. CXIV; Holtzapffel, vol. 5, "Principles and Practice of Ornamental or Complex Turning."

14. Victor Arwas, *Art Deco Sculpture: Chryselephantine Statuettes of the Twenties and Thirties,* New York: St. Martin's Press, 1975.
15. Yuzuru Okada, *Netsuke: A Miniature Art of Japan,* Tokyo: Japan Travel Bureau, 1951, p. 18.
16. Egerton Ryerson, *The Netsuke of Japan: Legends, History, Folklore, and Customs,* New York: A. S. Barnes & Co., 1958, p. 16.
17. Frederick Meinertzhagen, *The Art of the Netsuke Carver,* London: Routledge & Kegan Paul, 1956, listing 48, p. 62.
18. Miriam Kinsey, *Contemporary Netsuke,* Tokyo and Rutland, Vt.: Charles E. Tuttle, 1977, pp. 133, 140.
19. Meinertzhagen, p. 54.
20. Albert Brockhous, *Netsukes,* New York: Duffield & Co., 1924, p. 25.
21. *Ibid.,* pp. 74, 92–93.
22. Kinsey, p. 96.
23. Ryerson, p. 14.
24. Kinsey, p. 140.
25. Mary Louise O'Brien, *Netsuke: A Guide for Collectors,* Tokyo and Rutland, Vt.: Charles E. Tuttle, 1965.
26. Laufer, p. 76.
27. Kunz, pp. 118, 508 (L. Zick); 509 (Li Hsao-yu).
28. A detailed account of the operation is in Holtzapffel, vol. 4, "Principles and Practice of Hand or Simple Turning," pp. 426–31.
29. "Chinese Ivory Balls: From an Old *Book of Wonders,* 1895," *Hobbies,* vol. 53 (1948), no. 5, p. 20.
30. A fine completed production, about 5 inches (13 cm.) in diameter, is shown in Eastham, Pl. 34.
31. Kunz, p. 119, 509 (Li Hsao-yu).
32. *Ibid.,* p. 66.
33. Philippowich, Fig. 84.
34. Kunz, p. 71.
35. Holtzapffel, vol. 4, pp. 433–34, and shown in Fig. 574.
36. Philippowich, pp. 303–4; two draw-string models are shown in Figs. 229, 230.
37. *Ibid.,* Fig. 230a.
38. Maeda, cols. 779–84.
39. Flayderman, p. 3.
40. Ritchie, *Modern Ivory Carving,* pp. 37–39.
41. Shown and described in Flayderman, pp. 90–91.
42. Flayderman, pp. 82–84.
43. Ritchie, *Modern Ivory Carving,* p. 54.
44. Shown in Flayderman, p. 62.

2 BOOK ACCESSORIES
1. Beigbeder, Figs. 52, 53.
2. Alfred Maskell, Pl. 19 (left figure).
3. *Ibid.,* Pl. 19 (right figure).
4. *Ibid.,* Pl. 28 (lower figure).
5. *Catalogue of the Permanent and Loan Collections of the Jewish Museum, London,* London: Jewish Museum, 1974, Pl. LXVII (Figs. 177–179).

[214] Notes

3 CLOCKS
1. Shown in Philippowich, p. 273.
2. Ray, *Eskimo Art*, Fig. 139.
3. Rupert Gould, *Oddities*, New York: University Books, 1965, p. 197.

4 CLOTH AND CLOTHES MAKING
1. Therle Hughes, p. 156.
2. Beigbeder, Fig. 125.
3. Both are shown in Flayderman, pp. 224–25.
4. *Ibid.,* shown on p. 126.
5. Katherine M. McClinton, *The Complete Book of Small Antiques Collecting*, New York: Coward-McCann, 1965, pp. 213–18.
6. Shown in Flayderman, pp. 54, 232–33.
7. McClinton, p. 205.
8. Carl L. Crossman, *The China Trade*, Princeton, N. J.: Pyne Press, 1972, pp. 190–91.

5 CONTAINERS
1. Martial, *Epigrams*, trans. by Walter C. A. Ker, Cambridge, Mass.: Harvard University Press, 1961, vol. 2, Epigram LXXVII.
2. Laufer, pp. 74–75; the ivory covers are shown in Pls. IX, X.
3. Eastham, Pl. 17.
4. Kunz, pp. 122–25.
5. Therle Hughes, p. 156.
6. Flayderman; shown on p. 143, from a private collection later auctioned off.
7. Therle Hughes, pp. 17-18, 24.
8. McClinton, p. 196.
9. Shown in Flayderman, p. 203.
10. *Ibid.,* shown on p. 221.
11. Doris M. Green, "Canes or Walking Sticks," *Hobbies*, vol. 64 (1960), no. 11, pp. 56–57.
12. Alfred Maskell, Pls. 70, 44, and 46, respectively.
13. Thurston, p. 8.
14. Woodhouse, *Ivories*, p. 79.
15. Eastham, p. 43 and Pl. 31.
16. Martial, vol. 2, Epigram XII.
17. Beigbeder, Fig. 91; Carrà, p. 142 and Pl. 65; discussed extensively in Alfred Maskell, pp. 374–77, and shown in Pls. 63, 64.
18. A mid-nineteenth century example is shown in Woodhouse, *Ivories*, p. 116.
19. An example is shown in Kunz, facing p. 309.
20. Clare Le Corbeiller, *European and American Snuff Boxes: 1730–1830*, London: Batsford, 1966, pp. 1–17, 83–86; ivory examples are shown in Figs. 649–655, 661–662, 664, 670.
21. Eastham, p. 41; a decorated snuff saucer is shown in Pl. 29 (Fig. B).
22. Perry, p. 125; the ivory scoop-and-chute is shown on p. 29.
23. Charles P. Woodhouse, *The Victorian Collector's Handbook*, New York: St. Martin's Press, 1970, pp. 210–11.
24. Thurston, p. 8.
25. Eastham, p. 30.
26. Shown in Ritchie, *Modern Ivory Carving*, pp. 104–6.

27. Thurston, Pl. II, with description on p. 6.

28. J. Cameron, "Ebony Carving Inlaid with Ivory," *Journal of Indian Art*, vol. 4 (1892), no. 34, pp. 8–9.

29. Maeda, cols. 779–84.

30. *George A. Hearn Collection of Carved Ivories*, New York: Gillis Press, 1908; shown on pp. 213, 215.

31. Coomaraswamy, p. 176.

32. I. E. S. Edwards, *Tutankhamun: His Tomb and Its Treasures*, New York: Alfred Knopf, 1977; shown on pp. 242–43.

33. J. L. Kipling, "Indian Ivory Carving," *Journal of Indian Art*, vol. 1 (1885), no. 7, pp. 49–53.

34. Laufer, p. 69.

6 COSTUME ARTICLES

1. Kunz, p. 104.
2. Ellis, pp. 45–52.
3. Edward A. Alpers, *Ivory and Slaves in East Central Africa*, Los Angeles: University of California Press, 1975.
4. Carrà, p. 100.
5. Woodhouse, *Ivories*, p. 89.
6. Cox, Pl. 5; a nineteenth-century European buckle, delicately carved with floral designs, is shown in Woodhouse, *Ivories*, p. 106.
7. Shown in Flayderman, pp. 162–71.
8. Dorothy F. Brown, "Celluloid Buttons," *Hobbies*, vol. 65 (1960), no. 3, pp. 54–55; and Brown, "Carved Vegetable Ivory Buttons," *Hobbies*, vol. 85 (1980), no. 5, p. 98.
9. Bertha Betensley, *Antique Buttonhooks for Shoes, Gloves, and Clothing*, Westville, Ind.: Educator's Press, 1975.
10. Shown in Flayderman, pp. 193, 207.
11. Thurston, p. 8.
12. Kunz, p. 253.
13. Laufer, p. 69.
14. Kunz, p. 253.
15. Eastham, Pl. 28.
16. Carrà, p. 100.
17. Shown in Beigbeder, pp. 71–72.
18. Shown in Edwards, p. 167.
19. Shown in Flayderman, p. 212.
20. Labrets are shown in Ray, *Eskimo Art*, Figs. 4 and 5.
21. James Mackay, *Encyclopedia of Small Antiques*, New York: Harper & Row, 1975, pp. 171–72.
22. Shown in Flayderman, pp. 274–75.
23. Williamson, p. 19.
24. Ray, *Artists of the Tundra and the Sea*, p. 112.
25. Statius, *Silvae*, trans. by J. H. Mozley, Cambridge, Mass.: Harvard University Press, 1961, vol. 1, bk. I, pt. III, p. 43.
26. Shown in Cammann, "Carvings in Walrus Ivory," facing p. 8; an Eskimo wearing snow goggles is shown in Ray, *Eskimo Art*, Fig. 4.
27. Howard Carter, *The Tomb of Tut-ankh-amen*, New York: Cooper Square, 1963, vol. 1, Pl. LXX.

[216] *Notes*

28. Kurt Stein, *Canes and Walking Sticks,* York, Penn.: G. Shumway, 1974, pp. 103–6. An ivory handle example is shown in Fig. 129; an illustrated ad of a boy with cane, from *Harper's Bazar* [sic] of Nov. 24, 1883, is shown on p. 105.
29. Shown in Flayderman, pp. 206–7.

7 EDUCATIONAL IVORIES
1. Shown in Flayderman, pp. 240–41.
2. Shown in Tardy, p. 244.
3. Shown in Philippowich, p. 250.
4. *Ibid.,* shown on p. 251.
5. *Ibid.,* p. 248, shows Zick's female model; the male model is shown on p. 249, and the female model by an unknown sculptor, on p. 252.
6. *Ibid.,* p. 250.
7. Personal communication from Eugen Philippowich, 11 February 1980.
8. Woodhouse, *Ivories,* p. 56.
9. *Ibid.,* p. 66.
10. Shown in Flayderman, p. 227.

8 FOOD-RELATED ARTICLES
1. Flayderman, p. 167; one is shown on p. 193, but may be of whalebone.
2. *Ibid.,* shown on p. 193.
3. Laufer, pp. 8, 67.
4. Cookie molds are shown in Flayderman, p. 195.
5. Barnett, pp. 91–94.
6. Shown in Laufer, Pl. IX (Fig. 1).
7. Athenaeus, *The Deipnosophists, or the Banquet of the Learned,* London: Bohn, 1854, vol. 1, bk. IV, no. 3, p. 212.
8. Alfred Maskell, pp. 369–70, and shown in Pls. 60, 61.
9. A German or Dutch late-seventeenth-century tankard, elaborately carved, is shown in Woodhouse, *Ivories,* p. 28. Even more elaborate, with excess that precludes actual use, is the tankard shown in Beihoff, p. 69.
10. Shown in Flayderman, pp. 198–99.
11. *Ibid.,* shown on p. 201, including what appears to be a strainer spoon, as well as unusual combined fork-and-knife items.
12. Beigbeder, Fig. 78.
13. Charles T. Bailey, *Knives and Forks,* London and Boston: Medici Society, 1927, Fig. 22–37, 40–41.
14. Kunz, p. 438; see also Kipling.
15. Kunz, p. 231.
16. N. B. Nelson, "Ivory: Its Sources and Uses," *Popular Science Monthly,* vol. 51 (1897), pp. 534–39.
17. Wilson, p. 115.
18. "Knife Rests," *Hobbies,* vol. 71 (1966), no. 6, p. 28.
19. Shown in Flayderman, p. 190.
20. Cameron, pp. 8–9.
21. Flayderman, pp. 160–66; pp. 174–89 show examples made of whale-tooth ivory, walrus ivory, and whalebone; many examples are shown in Kunz, facing p. 482.
22. Shown in Flayderman, p. 193.
23. *Ibid.,* shown on p. 194.

Notes [217]

24. *Ibid.*, shown on p. 192.
25. *Ibid.*, shown on p. 255.
26. Fagg, pp. IX–XI.
27. Shown in Flayderman, p. 201.
28. Therle Hughes, Fig. 33.
29. Pieced ivory trays of great elegance are shown in *George A. Hearn Collection*, pp. 213, 215; of unspecified origin, both trays are 19 inches (48 cm.) in diameter.

9 FURNITURE

1. Kunz, p. 102.
2. Carter, vol. 3, Pl. XXXVI.
3. Discussed and shown in Edwards, pp. 210–11.
4. Shown in Kunz, facing p. 8.
5. Two luxuriously carved lampshades are shown in *George A. Hearn Collection*, p. 247.
6. Shown in Flayderman, p. 206.
7. Kunz, p. 14.
8. Alfred Maskell, p. 146, and shown in Pl. 18; also shown in Carrà, Pl. 30.
9. Kunz, pp. 33–34.
10. *Ibid.*, p. 292.
11. Both are shown in Alfred Maskell, Pl. 25.
12. Wyatt, vol. 2, Pl. CLVII; described in Thurston, p. 5.
13. Shown in Flayderman, p. 193.

10 GAMES

1. Except where otherwise indicated, most of the data concerning billiard balls is based on Kunz, pp. 244–49, 347.
2. T. A. Marchmay, "Making Billiard Balls from Ivory," *Scientific American Monthly*, vol. 3 (1921), pp. 316–18.
3. Holtzapffel, vol. IV, p. 408.
4. "Wanted—Substitute for Ivory for Billiard Balls," *Illustrated World*, vol. 32 (1920), p. 926.
5. Barnett, pp. 123–24.
6. Loud, Pls. 1–63.
7. Examples are shown in Alfred Maskell, Pl. 69.
8. *Ibid.*, Pl. 68.
9. Ritchie, *Modern Ivory Carving*, p. 126.
10. *Ibid.*, pp. 78, 84.
11. Woodhouse, *Ivories*, p. 63.
12. Carrà, Pl. 64.
13. Therle Hughes, Fig. 31a.
14. Ray, *Eskimo Art*, pp. 43–44.
15. Homer, *The Odyssey*, trans. by William Cowper, New York: Dutton & Co., 1910, bk. I, p. 4.
16. Martial, vol. 2, Epigram XIV.
17. Ray, *Artists of the Tundra and the Sea*, p. 113 and Fig. 28.
18. Carson Ritchie, *Scrimshaw*, New York: Sterling, 1972, pp. 37–38, shown on p. 37.
19. Carter, vol. 3, Pls. XLII, LXXV.
20. Flinders Petrie, *Objects of Daily Use*, London: British School of Archeology in Egypt, 1927, pp. 51–57 and Pls. XLVII, XLVIII.

[218] *Notes*

21. Loud, Pl. 47.
22. Edwards, pp. 208-9.
23. Crossman, p. 194.
24. Laufer, p. 77.
25. Kunz, p. 101.
26. Shown in Flayderman, p. 254.

11 HEALTH ARTICLES
1. Holiday, pp. 564-74.
2. Kunz, p. 320.
3. Beihoff, p. 85; in reply to the present writer's inquiry, Norbert Beihoff said he no longer had the reference for that advertisement.
4. Holtzapffel, vol. I, p. 139.
5. Kunz, p. 292.
6. Shown in Flayderman, pp. 132-33, including a rare cane made from a walrus backbone; see also Stein.
7. Eastham, pp. 15-17
8. Kunz, p. 292.
9. *Ibid.*, p. 438.
10. "Wanted—Substitute for Ivory for Billiard Balls," *Illustrated World.*
11. Kunz, p. 186 (quoting from Edward Topsell, *History of Four-footed Beasts,* London, 1658).
12. Kunz, p. 84.
13. *Ibid.*, pp. 409-10.
14. *Ibid.*, p. 265.
15. *Ibid.*, p. 318.
16. Cammann, "Account of Hornbill Ivory."
17. Lucian, "Remarks Addressed to an Illiterate Book Fancier," *The Works of Lucian,* vol. 3, trans. by H. W. Fowler and F. G. Fowler, London: Oxford University Press, 1949, p. 278.
18. Martial, vol. 2, Epigram LXXVIII.
19. Barnett, pp. 123-24.
20. Laufer, p. 60.
21. Kunz, p. 253.
22. Mackay, pp. 300-301.

12 HOUSE ACCESSORIES
1. Williamson, pp. 30-31, 38-39.
2. *Ibid.*
3. Thurston, Pl. V.
4. Ritchie, *Modern Ivory Carving,* p. 29.
5. Shown in Flayderman, p. 205, from a private collection (now auctioned off).
6. Williamson, pp. 30-31, 38-39.

13 MUSICAL INSTRUMENTS
1. Shown in Kunz, facing p. 9.
2. Coomaraswamy, p. 177.
3. Barnett, pp. 123-24.
4. Alfred Maskell, pp. 460-63 and Pl. 80; Carrà, p. 128 and Pl. 58.

5. Shown in Flayderman, pp. 136–37.
6. Kunz, pp. 250–54.
7. *Ibid.*
8. F. H. Andrews, "The Elephant in Industry and Art: Part 2," *Journal of Indian Art*, vol. 10 (1904), no. 88, pp. 55–64.
9. Ryerson, p. 12.
10. Andrews, pp. 55–64.

14 PERSONAL COMFORT ARTICLES
1. Martial, vol. 2, Epigram LXXXIII.
2. Woodhouse, *The Victorian Collector's Handbook*, p. 193.
3. Carol Dorrington-Ward, ed., *Fans from the East*, New York: Viking Press, 1978, pp. 27, 38.
4. G. W. Rhead, *History of the Fan*, London: Kegan Paul & Co., 1910, pp. 136–37.
5. Dorrington-Ward, Fig. 48.
6. Nancy Armstrong, *A Collector's History of Fans*, New York: Clarkson Potter, 1974, p. 13.
7. Dorrington-Ward, pp. 37–38.
8. Armstrong, *A Collector's History of Fans*, p. 152.
9. Rhead, p. 174.
10. Armstrong, *The Book of Fans*; Rhead, *History of the Fan*.
11. Anacreon, "Poems," bk. IV, nos. 96–97, in *Lyra Graeca*, vol. II, trans. and ed. by J. M. Edmonds, Cambridge, Mass.: Harvard University Press, 1964, p. 189.
12. T. S. Crawford, *A History of the Umbrella*, New York: Taplinger, 1970, chap. 6.

15 PERSONAL GROOMING ARTICLES
1. Kunz, p. 265.
2. *Ibid.*, p. 253.
3. An ivory comb from the period 5500–4000 B.C. can be seen in Michael A. Hoffman, *Egypt before the Pharaohs*, New York: Alfred Knopf, 1979, p. 139.
4. Shown in Kunz, facing p. 13.
5. Hispanic Society of America, *Early Engraved Ivories in the Collection*, New York: Hispanic Society of America, 1928, pp. 1–11, 52–66.
6. Carrà, p. 86, and Pl. 39.
7. Williamson, p. 11.
8. Geoffrey Wills, *Ivory*, New Jersey: A. S. Barnes, 1969, Pl. 6.
9. Beigbeder, Fig. 66. Other decorative ivory combs in the same museum—fourteenth to sixteenth-century Italian and English carvings—are shown in Alfred Maskell, Pl. 50.
10. Havell, pp. 20–21.
11. Kunz, p. 253.
12. Havell, pp. 20–21.
13. Cecil L. Burns, "A Monograph on Ivory Carving," *Journal of Indian Art*, vol. 9 (1901), no. 75, pp. 53–56.
14. Examples are shown in Cox, Pl. 4.
15. Beihoff, p. 35.
16. Kunz, p. 253.
17. Examples are shown in Beigbeder, Pl. 24; Carrà, Pl. 59; and Alfred Maskell, Pls. 48–49.

[220] Notes

18. Ritchie, *Ivory Carving*, Fig. 7.
19. Coomaraswamy, p. 177.

16 PUZZLES
1. One is shown in Crossman, p. 193.

17 RELIGIOUS ARTICLES
1. Perhaps the finest example is seen in "The Descent from the Cross" that is shown in the frontispiece of Alfred Maskell's *Ivories*.
2. Shown in Cammann, "Carvings in Walrus Ivory," facing p. 22.
3. Alfred Maskell, Pls. 19, 20, 28.
4. Williamson, p. 183.
5. Tardy, pp. 308–9.
6. Beihoff, p. 84.
7. Alfred Maskell, p. 337 and Pl. 54; Maskell applies the term "morbid" to such supplementation of ivory art.
8. Shown in John Beckwith, *Ivory Carving in Early Medieval England*, Greenwich, Conn.: New York Graphic Society, 1972, Figs. 99–102.
9. An ivory crucifix on a wooden cross, seventeenth century European, is shown in Woodhouse, *Ivories*, p. 21; another, around 1800, is shown on p. 91.
10. Two spherical diptychs are shown in *George A. Hearn Collection*, p. 43.
11. Many are shown in Alfred Maskell, Pls. 29–31; Beigbeder, Figs. 21, 22, 29; and Carrà, Pl. 32.
12. Alfred Maskell, Pl. 55.
13. Beigbeder, Figs. 52–53.
14. William Maskell, p. 91.
15. Some are shown in Alfred Maskell, Pls. 41–42.
16. Some are shown in: Alfred Maskell, Pl. 40 (Fig. 5); Pelka, p. 133; Kunz, facing p. 68.
17. Dorrington-Ward, Pl. 20; comments are on pp. 25, 62, 66.
18. Kunz, p. 99.
19. Some are shown in Beckwith, Figs. 98, 128–29, 132–33; also in Alfred Maskell, Pl. 43.
20. Shown in Alfred Maskell, Pl. 24 (Fig. 2).
21. Beigbeder, p. 65; paxes are shown in Figs. 68–69.
22. Pyxes are shown in Pelka, pp. 30, 42, 56–57; one is shown in Alfred Maskell, Pl. 12 (Fig. 5).
23. Carrà, Pl. 50.
24. Alfred Maskell, p. 347.
25. Beihoff, p. 91.
26. Petrie, pp. 39–43 and Pls. xxxv–xxxvii.
27. Kunz, p. 17.
28. *Ibid.*, p. 292.
29. *Ibid.*, pp. 409–10.
30. Havell, pp. 20–21.
31. Ray, *Eskimo Art*, pp. 17–20.
32. Woodhouse, *Ivories*, p. 107.
33. Two ivory fetishes with carved faces, from Africa, are shown in an article by Charles Miles, "Fetishes," *Hobbies*, vol. 62 (1958), no. 12, pp. 112–14, Fig. 1.

34. Cammann, "Carvings in Walrus Ivory."
35. William Maskell, p. 50; Carrà, p. 86 and Pl. 40.
36. Williamson, pp. 179–80.
37. Dalton, p. xxiv.
38. Williamson, pp. 156–57; ivory holy vessels are shown in Pelka, p. 138.

18 SCIENTIFIC INSTRUMENTS
1. Coomaraswamy, Fig. 133.
2. Martha Hill Hommel, "Keeping Count," *Hobbies*, vol. 61 (1956), no. 4, p. 45.
3. Shown in *The Complete Encyclopedia of Antiques*, New York: Hawthorn Books, 1962, Pl. 438 (Fig. C).
4. Ritchie, *Ivory Carving*, p. 35.
5. Ritchie, *Modern Ivory Carving*, p. 41.
6. *The Complete Encyclopedia of Antiques*, p. 1196 and Pl. 436.

19 SMOKING ARTICLES
1. Shown in Cammann, "Carvings in Walrus Ivory," facing p. 8.
2. Ray, *Artists of the Tundra and Sea*, Figs. 47–48.
3. *Ibid.*, Figs. 49–50; other Eskimo pipes, carved and painted elaborately, are shown in Ray's later treatise, *Eskimo Art*, Figs. 242–45.
4. Woodhouse, *The Victorian Collector's Handbook*, p. 208.
5. Shown in Flayderman, p. 141.
6. Opium pipes are shown in Eastham, Pl. 22 (Fig. A).
7. Cox, Pl. 44.
8. Opium jars are shown in Eastham, Pl. 24 (Figs. D, E).

20 TOYS
1. Bosticco, Pl. 236.
2. Flayderman, pp. 183, 185.
3. Shown in Flayderman, p. 238.
4. Ray, *Eskimo Art*, Figs. 44–46.
5. Shown in an article by Charles Miles, "The Eskimo Culture Area: Part II," *Hobbies*, vol. 70 (1965), no. 9, pp. 112–14.
6. Burns, pp. 53–56.

21 WEAPONS AND HUNTING ARTICLES
1. Alfred Maskell, Pl. 82 (lower center fig.).
2. Beihoff, p. 7.
3. Ray, *Eskimo Art*, Figs. 27, 232.
4. Laufer, pp. 52–53.
5. Cammann, "Carvings in Walrus Ivory."
6. Laufer, p. 7.
7. Flayderman, p. 21.
8. Cox, Pl. 3.
9. Kunz, p. 292.
10. Carrà, p. 136 and Pl. 60.
11. Shown in Flayderman, p. 130.
12. Metal fishhooks were preferable, but metal came late to that part of the world.
13. Various forms of such ivory lures are shown in Ray, *Eskimo Art*, Fig. 41.

14. Ray, *Artists of the Tundra and the Sea,* Figs. 15–16, 19, 21.
15. *Ibid.,* Fig. 27.
16. Ray, *Eskimo Art,* Figs. 15, 16.
17. Carrà, Pl. 42.
18. *Ibid.,* Pl. 41.
19. Alfred Maskell, p. 314 and Pl. 51 (Fig. 2).
20. Examples are shown in Alfred Maskell, Pl. 82; a boar's tusk model is shown in Beihoff, p. 65.
21. Williamson, p. 46.
22. Shown in Flayderman, p. 131.
23. Carter, vol. 1, Pl. LXXII.

22 WRITING AND PAINTING INSTRUMENTS

1. Carter, vol. 3, p. 81 and Pl. XXII.
2. Edwards, shown with comments on p. 181.
3. Kunz, p. 90.
4. The top and bottom views of an arm rest are shown in Eastham, Pl. 21.
5. Barnett, p. 16.
6. Shown in Flayderman, p. 139.
7. Hoffman, p. 274.
8. Laufer, pp. 8, 72–73.
9. Eastham, pp. 42–43.
10. *Ibid.,* pp. 2–3.
11. Shown in Laufer, Pls. VI and VII, are two ivory paintbrush holders with scenes carved in relief, Chien Lung period (A.D. 1736–95); one with carved human figures is shown in Eastham, Pl. 23 (Fig. A).
12. Kunz, p. 10.
13. Carter, vol. 3, Pl. XXII.
14. *Ibid.,* Pl. LXVI (Fig. A).
15. Barnett, pp. 123–24.
16. Beckwith, Figs. 95–97.
17. Maeda, cols. 779–84.
18. Eastham, pp. 28, 42.
19. Ryerson, p. 14.
20. Charles Miles, "Toys," *Hobbies,* vol. 81 (1976), no. 4, pp. 142–45; a snow knife is shown on p. 143.
21. Burns, pp. 53–56.
22. Many are shown in Alfred Maskell, Pls. 5–8.
23. William Maskell, p. 110.

23 MISCELLANEOUS

1. Joseph Mayer, *Catalogue of the Fejervary Ivories in the Museum of Joseph Mayer,* Liverpool: Marples, 1856, p. 34.
2. Maeda, cols. 779–84.
3. Laufer, p. 8.
4. Philippowich, p. 303.
5. Laufer, p. 9, and shown in Pl. I (Fig. 1); Cox, Pl. 2.
6. Wilson and Ayerst, p. 115.
7. Letter from Peter Ayerst, 29 July 1980.

Notes [223]

8. Shown in Flayderman, p. 119; others are shown on p. 120.
9. Ray, *Artists of the Tundra and the Sea,* Fig. 35; and Ray, *Eskimo Art,* Fig. 40.
10. Havell, pp. 20–21.
11. Stubbs, pp. 41–45.
12. Williamson, p. 106.
13. Finely sculpted oliphants are shown in Alfred Maskell, Pl. 51.
14. Kunz, p. 99.
15. The palanquin of the Maharajah of Travancore State, made around the 1840s, is shown in Thurston, Pl. I (top).
16. Kunz, p. 228.
17. An example is shown in Flayderman, p. 125, from a private collection (now auctioned off).
18. Kunz, pp. 27, 292.
19. Shown in Kunz, facing p. 123.
20. Robert Webster, "Amber, Jet, and Ivory," *The Gemmologist,* vol. 27 (1958), no. 321, pp. 65–72; the ivory mace is shown on p. 71.
21. Burns, pp. 53–56.
22. Shown in Penniman, Pl. IX (Fig. 2).
23. Alfred Maskell, Pl. 67; other ivory graters are shown in Woodhouse, *Ivories,* p. 32.
24. Flayderman, p. 150.
25. Hispanic Society of America, pp. 1–11, 106–7.
26. Ray, *Eskimo Art,* Fig. 5.
27. Shown in the process of being used, in Ray, *Artists of the Tundra and the Sea,* Fig. 93; decorated drill bows are shown in Ray, *Eskimo Art,* Fig. 2.
28. Shown in Flayderman, p. 123.
29. Shown in Cammann, "Carvings in Walrus Ivory," facing p. 8.
30. Shown in Flayderman, p. 204.

BIBLIOGRAPHY

Albes, Edward. "Tagua—Vegetable Ivory," *Bulletin of the Pan American Union*, vol. 37 (1913), pp. 192–208
Alpers, Edward A. *Ivory and Slaves in East Central Africa*. Los Angeles: University of California Press, 1975
Anacreon. "Poems." In *Lyra Graeca*, vol. II, ed. and trans. by J. M. Edmonds. Cambridge, Mass.: Harvard University Press, 1964
Andrews, F. H. "The Elephant in Industry and Art: Part 2," *Journal of Indian Art*, vol. 10 (1904), no. 88, pp. 55–64
Armstrong, Nancy. *A Collector's History of Fans*. New York: Clarkson Potter, 1974
———. *The Book of Fans*. New York: Mayflower Books, 1978
"Artificial Ivory," *Scientific American*, vol. 64 (1891), p. 73
Arwas, Victor. *Art Deco Sculpture: Chryselephantine Statuettes of the Twenties and Thirties*. New York: St. Martin's Press, 1975
Athenaeus. *The Deipnosophists, or the Banquet of the Learned*. London: Bohn, 1854
Baer, N. S., and others. "The Effect of High Temperature on Ivory," *Studies in Conservation*, vol. 16 (1971), no. 1, pp. 1–8
Bailey, Charles T. *Knives and Forks*. London and Boston: Medici Society, 1927
Barnett, R. D. *A Catalogue of the Nimrud Ivories, with Other Examples of Ancient Near Eastern Ivories*. London: British Museum, 1957
Beckwith, John. *Ivory Carving in Early Medieval England*. Greenwich, Conn.: New York Graphic Society, 1972
Beigbeder, O. *Ivory*. New York: Putnam's Sons, 1965
Beihoff, Norbert. *Ivory Sculpture through the Ages*. Milwaukee: Public Museum, 1961
Betensley, Bertha. *Antique Buttonhooks for Shoes, Gloves, and Clothing*. Westville, Ind.: Educator's Press, 1975
Bosticco, Sergio. "Ivory and Bone Carving." In *Encyclopedia of World Art*, vol. 8. New York: McGraw-Hill, 1963
Botsford, Virginia. "Celluloid," *Collectibles Monthly*, May, June, July, 1978
Brockhous, Albert. *Netsukes*. New York: Duffield & Co., 1924
Brode, Heinrich. *Tippoo Tib: The Story of His Career in Central Africa*. London: Edward Arnold, 1907
Brown, Dorothy F. "Celluloid Buttons," *Hobbies*, vol. 65 (1960), no. 3, pp. 54–55

———. "Carved Vegetable Ivory Buttons," *Hobbies,* vol. 85 (1980), no. 5, p. 98
Burns, Cecil L. "A Monograph on Ivory Carving," *Journal of Indian Art,* vol. 9 (1901), no. 75, pp. 53–56
Bushell, Raymond. *The Netsuke Handbook of Ueda Reikichi.* Tokyo and Rutland, Vt.: Charles E. Tuttle, 1961
Cameron, J. "Ebony Carving Inlaid with Ivory," *Journal of Indian Art,* vol. 4 (1892), no. 34, pp. 8–9
Cammann, Schuyler. "Jewelry from a Bird," *Antiques,* vol. 66 (1954), July, pp. 34–36
———. "Ivory and Bone Carving." In *Encyclopedia of World Art,* vol. 8. New York: McGraw-Hill, 1963
———. "Account of Hornbill Ivory," *Pennsylvania University Museum Bulletin,* vol. 15 (1950), no. 4, pp. 19–47
———. "Carvings in Walrus Ivory," *Pennsylvania University Museum Bulletin,* vol. 18 (1954), no. 3, pp. 2–31
Carrà, Massimo. *Ivories of the West.* New York: Hamlyn, 1970
Carter, Howard. *The Tomb of Tut-ankh-amen.* New York: Cooper Square, 1963
Catalogue of the Permanent and Loan Collections of the Jewish Museum, London. London: Jewish Museum, 1974
"Chinese Ivory Balls: from an Old *Book of Wonders,* 1895," *Hobbies,* vol. 53 (1948), no. 5, p. 20
Complete Encyclopedia of Antiques. New York: Hawthorn Books, 1962
Coomaraswamy, Ananda K. *The Arts and Crafts of India.* New York: Farrar Straus, 1964
Cox, Warren E. *Chinese Ivory Sculpture.* New York: Crown, 1946
Crawford, T. S. *A History of the Umbrella.* New York: Taplinger, 1970
Crossman, Carl L. *The China Trade.* Princeton, N.J.: Pyne Press, 1972
Dalton, O. M. *Catalogue of the Ivory Carvings of the Christian Era with Examples of Mohammedan Art and Carvings in Bone, in the British Museum.* London: British Museum, 1909
Dorrington-Ward, Carol, ed. *Fans from the East.* New York: Viking Press, 1978
"Doum Nuts of Commerce," *Scientific American,* vol. 116 (1917), p. 129
Eastham, Barry C. *Chinese Art Ivory.* Ann Arbor, Mich.: Ars Ceramica, 1976
Edwards, I. E. S. *Tutankhamun: His Tomb and Its Treasures.* New York: Alfred Knopf, 1977
Ellis, T. P. "Ivory Carving in the Punjab," *Journal of Indian Art,* vol. 9 (1901), no. 75, pp. 45–52
Fagg, William B. *Afro-Portuguese Ivories.* London: Batchworth Press, 1959
Flayderman, E. Norman. *Scrimshaw and Scrimshanders: Whales and Whalemen.* New Milford, Conn.: N. Flayderman & Co., 1972
George A. Hearn Collection of Carved Ivories. New York: Gillis Press, 1908
Gould, Rupert. *Oddities.* New York: University Books, 1965
Great Exhibition of the Works of Industry of All Nations, 1851. Official Descriptive and Illustrated Catalogue. London: Spicer Brothers, 1851
Green, Doris M. "Canes or Walking Sticks," *Hobbies,* vol. 64 (1960), no. 11, pp. 56–57
Havell, E. B. "Ivory Carving in Madras," *Journal of Indian Art,* vol. 2 (1888), pp. 20–21
Hispanic Society of America. *Early Engraved Ivories in the Collection.* New York: Hispanic Society of America, 1928
Hoffman, Michael A. *Egypt before the Pharaohs.* New York: Alfred Knopf, 1979
Holiday, Clayton. "The Conservation of Ivory and Bone," *Bulletin of the South African Museums Association,* vol. 9 (1971), no. 16, pp. 564–74

Holtzapffel, Charles. *Turning and Mechanical Manipulation,* 6 vols., 2nd ed. London: Holtzapffel & Co., 1846
Homer. *The Odyssey,* trans. by William Cowper. New York: Dutton & Co., 1910
Hommel, Martha Hill. "Keeping Count," *Hobbies,* vol. 61 (1956), no. 4, p. 45
Hughes, G. B. "Mechanical Carving Machines," *Country Life,* Sept. 23, 1954, pp. 980–81
Hughes, Therle. *Small Antiques for the Collector.* New York: Macmillan Co., 1964
Kinsey, Miriam. *Contemporary Netsuke.* Tokyo and Rutland, Vt.: Charles E. Tuttle, 1977
Kipling, J. L. "Indian Ivory Carving," *Journal of Indian Art,* vol. 1 (1885), no. 7, pp. 49–53
"Knife Rests," *Hobbies,* vol. 71 (1966), no. 6, p. 28
Kunz, George F. *Ivory and the Elephant: in Art, in Archeology, and in Science.* New York: Doubleday, 1916
Laufer, Berthold. *Ivory in China.* Chicago: Field Museum of Natural History, 1925
Le Corbeiller, Clare. *European and American Snuff Boxes: 1730-1830.* London: Batsford, 1966
Longhurst, Margaret. *English Ivories.* London: Putnam's Sons, 1926
Loud, Gordon. *The Megiddo Ivories.* Chicago: University of Chicago Press, 1939
Lucian. "Remarks Addressed to an Illiterate Book Fancier." In *The Works of Lucian,* vol. 3, trans. by H. W. Fowler and F. G. Fowler. London: Oxford University Press, 1949
McClinton, Katherine M. *The Complete Book of Small Antiques Collecting.* New York: Coward-McCann, 1965
Mackay, James. *Encyclopedia of Small Antiques.* New York: Harper & Row, 1975
Maeda, Taiji. "Ivory and Bone Carving." In *Encyclopedia of World Art,* vol. 8. New York: McGraw-Hill, 1963
Malley, Richard C. *Graven by the Fishermen Themselves: Scrimshaw in Mystic Seaport Museum.* Mystic, Conn.: Mystic Seaport Museum, 1983
Marchmay, T. A., "Making Billiard Balls from Ivory," *Scientific American Monthly,* vol. 3 (1921), pp. 316–18
Martial. *Epigrams.* Cambridge, Mass.: Harvard University Press, 1961
Maskell, Alfred. *Ivories.* London: Methuen, 1905. Rpt. ed., Tokyo and Rutland, Vt.: Charles E. Tuttle, 1966
Maskell, William. *Ivories Ancient and Medieval.* London: Chapman & Hall, 1870
Mayer, Joseph. *Catalogue of the Fejervary Ivories in the Museum of Joseph Mayer.* Liverpool: Marples, 1856
Meinertzhagen, Frederick. *The Art of the Netsuke Carver.* London: Routledge & Kegan Paul, 1956
Miles, Charles. "Fetishes," *Hobbies,* vol. 62 (1958), no. 12, pp. 112–14
―――. "The Eskimo Culture Area: Part II," *Hobbies,* vol. 70 (1965), no. 9, pp. 112–14
―――. "Toys," *Hobbies,* vol. 81 (1976), no. 4, pp. 142–45
Mills, John F. *The Care of Antiques.* New York: Hastings House, 1964
Moore, Ernst D. *Ivory: Scourge of Africa.* New York: Harper, 1931
Nelson, N. B. "Ivory: It's Sources and Uses," *Popular Science Monthly,* vol. 51 (1897), pp. 534–39
O'Brien, Mary Louise. *Netsuke: A Guide for Collectors.* Tokyo and Rutland, Vt.: Charles E. Tuttle, 1965
Okada, Yuzuru. *Netsuke: A Miniature Art of Japan.* Tokyo: Japan Travel Bureau, 1951
Pelka, Otto. *Elfenbein.* Berlin: Schmidt & Co., 1920
Penniman, Thomas K. *Pictures of Ivory and Other Animal Teeth, Bone, and Antler: with a*

Brief Commentary on Their Use in Identification. Oxford: Oxford University Press, 1952
Perry, Lilla. *Chinese Snuff Bottles.* Tokyo and Rutland, Vt.: Charles E. Tuttle, 1960
Petrie, Flinders. *Objects of Daily Use.* London: British School of Archeology in Egypt, 1927
Philippowich, Eugen. *Elfenbein.* Braunschweig, Germany: Klinkhardt & Biermann, 1961
Pratt, Julius. *Centennial of Meriden,* June 10–16, 1901
Ray, Dorothy Jean. *Artists of the Tundra and the Sea.* Seattle: University of Washington Press, 1961
_____. *Eskimo Art: Tradition and Innovation in North Alaska.* Seattle: University of Washington Press, 1977
Rhead, G. W. *History of the Fan.* London: Kegan Paul & Co., 1910
Ritchie, Carson. *Ivory Carving.* London: Arthur Barker, 1969
_____. *Modern Ivory Carving.* New Jersey: A. S. Barnes, 1972
_____. *Scrimshaw.* New York: Sterling, 1972
Ryerson, Egerton. *The Netsuke of Japan: Legends, History, Folklore, and Customs.* New York: A. S. Barnes, 1958
Savage, George. *The Antique Collector's Handbook.* London: Barrie & Rockliff, 1959
Solis, Acosta. *La Tagua.* Quito, Ecuador: Publicaciones Cientificas, 1944
South Kensington Museum. *A Description of the Ivories Ancient and Medieval in the South Kensington Museum, with a Preface by William Maskell.* London: Chapman & Hall, 1872
Spielman, M. H. "Art Forgeries and Counterfeits: Ivories," *Magazine of Art,* vol. 27 (1903), pp. 549–50
Statius. *Silvae.* Cambridge, Mass.: Harvard University Press, 1961
Stein, Kurt. *Canes and Walking Sticks.* York, Penn.: G. Shumway, 1974
Stubbs, L. M. "Ivory Carving in the Northwest Provinces and Oudh," *Journal of Indian Art,* vol. 9 (1901), no. 75, pp. 41–45
Tardy and others. *Les Ivoires.* Paris: Tardy, 1966.
Thurston, Edgar. *Monograph on the Ivory Carving Industry of Southern India.* Madras: Government Press, 1901
"Wanted—Substitute for Ivory Billiard Balls," *Illustrated World,* vol. 32 (1920), p. 926
Webster, Robert. "Amber, Jet, and Ivory," *The Gemmologist,* vol. 27 (1958), no. 321, pp. 65–72
_____. "Ivory, Bone, and Horn," *The Gemmologist,* vol. 27 (1958), no. 322, pp. 91–98
_____. "Vegetable Ivory and Tortoise Shell," *The Gemmologist,* vol. 27 (1958), no. 323, pp. 103–7
Williamson, George C. *The Book of Ivory.* London: Frederick Muller, 1938
Wills, Geoffrey. *Ivory.* New Jersey: A. S. Barnes, 1969
Wilson, Derek, and Ayerst, Peter. *White Gold: The Story of African Ivory.* New York: Taplinger, 1976
Woodhouse, Charles P. *Ivories: A History and Guide.* New York: Van Nostrand Reinhold, 1976
_____. *The Victorian Collector's Handbook.* New York: St. Martin's Press, 1970
Wyatt, M. Digby. *Industrial Arts of the Nineteenth Century at the Great Exhibition, 1851.* London: Day & Son, 1853

INDEX

abstinence plaques, 31, 167
admission tickets, 195
adzes, 201
Africa: ivory and slave trade in, 22, 23–25; ivory carving in, 25; uses of ivory in, 96, 135, 138, 176
Alessandri, M., 46
alphabet sets and tablets, 111
American Museum of Natural History, 37, 84, 85, 180
American whaler seaman, specific items made by: alphabet sets, 111; ball-and-cup toy, 181; candle holders, 98; dagger handles, 184; duster handles, 202; inkwells, 188; jewelry trees, 108; knobs, 135; mallets, 202; neckerchief slides, 109; pointers, 113; rope-bedspring tighteners, 120; scrimshaw, 27; seals, 193; seam rubbers, 94; utensils, 113, 114, 116, 117, 118; whips, 187
amulets, 132, 175–76
Anatomical Institute (Basel), 112
anatomical models, 22, 111–12
Anchorage Museum, 92
Angermair, Christoph, 98, 99
animal cages, 20, 96–97
animal-hide scrapers, 92, 93
anklets, 16, 104
apple corers, 113
Arab trade in ivory, 24–25
architectural compasses, 177

architectural models, 112
armlets, 16, 103
armor, 17, 183–84
arm rests, 17
arrowheads, 16, 183
arrow-shaft straighteners, 183
Art Deco sculpture, 80–81
artificial flowers, ivory-palm nut, 38
artificial ivory, 39–40, 57
artificial teeth, 130–31
artistic ivories, 17, 51–52, 77–89
Asian Art Museum, 85
aspergillum, 171
Assyria, 16, 77; specific items made in: bowls, 114; containers, 96; fly whisks, 158; inkwells, 188; spoons, 115; writing styli, 193
Athena, statue of, 19, 43
atomizers, 164
awls, 92, 201. See also bodkins

babirusa, tusk of, 34
back scratchers, 17, 140
baleen, 36; specific items of: bows, 184; bow-tips, 184; canes, 131; caps and hats, 107; shoehorns, 110; swifts, 95; tongue depressors, 133; toys, 181; whips, 187
baleen whales, 36
ball-and-cup toy, 181
balls. See roulette-wheel balls; billiard balls

Index

bamboo ivory, 29
bangles, 18, 103–4
Bargello Museum, 125
baskets, 20, 103; bread, 114
basting-thread removal pins, 93
batons, 17, 195; conductors', 137
Bavarian National Museum, 99
beach ivory, 30
beads, 109
bedposts, 119
beds, 16, 119
belt knives, 199
betel nut, 39
biblical references to ivory, 16, 119, 121, 135, 137
billiard balls, 26, 29, 39–42, 122–24
billiard cues, 124
billy clubs, 187
binoculars, 178
birdcages, 96–97
Birmingham City Museum, 160
blemish removers, 161
boar, tusk of, 34, 132
boat fittings, 26, 195
boats, miniature, 17, 79
bodkins, 16, 93
bolas, 184
bone, 16, 35–36, 57, 58, 59; specific items of: flutes, 137; games, 126, 128; neckerchief slides, 109; razor handles, 165; sewing needles, 92; tools, 92, 93, 94; toothbrushes, 133; weapons, 183; window-shade pulls, 121. *See also* porpoise; whalebone
book accessories: bookmarks, 90; covers, 18, 21, 43, 45, 90, 167; bookstands, 18, 90–91
books of ivory, 90
boot hooks, 110
Borelli, Antonio, 23
Borneo, 37
bottle stoppers, ivory-palm nut, 38
bow drills, 201
bowls, 16, 114
bows, 184
bow-tips, 17, 184
boxes, 16, 18, 20, 21, 40, 103; cigar, 179; cold cream, 161; face-powder, 162; glove, 106–7; glove-powder, 107; hairpin, 163; handkerchief, 99; incense, 17, 99; jewelry, 17, 19, 108; pill, 18, 133; pin, 100; scent, 17, 100; seal, 17, 193; seal-pigment, 193; stamp, 101; tobacco, 180; tooth-cleaning powder, 134
bracelets, 103–4
bread baskets, 114
British Museum, exhibits in: Assyrian ivories, 16; chairs, 119; clappers, 137; games and accessories, 124, 127; memento mori, 79; mirror cases, 164; models of Parthenon, 112; netsuke, 85; pen case, 192; plaque, 90; portrait medallions, 78; religious articles, 169, 171, 172; seals, 193; shaker instrument, 139; tankards, 114; trinity rings, 22; weapons, 184, 185
brooches, 21, 37, 104, 108
brow-bands, 104
brushes: clothes, 106; face-powder, 162; hat, 107; painting, 190; shaving, 165; writing, 190. *See also* hairbrush backs and handles; toothbrush handles
buckles, 37, 104
Buddhism, 173
burnishers, 187
business cards, 187
busks, 105
busts, 78
buttonhooks, 105–6
buttons, 36, 38, 39, 105; collar, 46, 105
Byzantine style, 20

cabinets, 119
cages. *See* animal cages; birdcages
calculators, 177–78
calling-card cases, 17, 18, 97–98
candelabra, 98
candle holders, 98
canes, 36, 131; as containers, 98. *See also* walking sticks
caps, 107
carriage wheels, baleen, 36
Carriera, Rosalba, 43
cases, 19, 98–99; calling-card, 17, 18, 97–98; cigarette, 179; clock, 23, 40, 91–92; envelope, 18, 99; glove, 18; handkerchief, 18; incense, 99; jewelry,

Index [231]

18, 21; medical instrument, 132–33; medicine, 133; mirror, 21, 22, 164; money, 18, 99; needle, 18, 38, 94; paintbrush, 17, 190; pen, 16, 191–92; pipe, 18, 180; spill, 101; stamp, 18; thimble, 95; toothpick, 134; for writing brushes, 190
caskets for valuables, 98
castanets, 136–37
Cellonite, 40
Celluloid, 40, 100, 106, 178
Ceylon, 18, 102, 137, 164, 184, 192
chairs, 16, 20, 119; coronation, 33
chalices, 21, 114
chariots, 17, 20, 195
charms, 26, 175–76
checkers, 38, 124
chessboards, 18, 125–26
chess pieces and sets, 17, 18, 21, 23, 38, 124–25
chests, 98, 103; jewelry, 108
Cheverton, Benjamin, 23, 45, 46
China, 16–18, 31–32, 33, 37, 43; ivory as food in, 116; ivory as medicine in, 131–32; ivory-incense burning in, 172; specific items made in: alphabet sets, 111; bow tips, 184; boxes and cases, 97–98, 99, 100, 108, 193; buckles, 104; buttons, 105; cages, 96–97; cigarette holders, 179; doctor's lady, 17, 131; fans, 140, 157, 158; flutes, 137; games and accessories, 125, 128; grooming and comfort articles, 140, 161, 163; hat stands, 107; incense burners, 99; interior carvings, 87; opium pipes, 180; paintbrush cups, 190; palanquins, 198; picture frames, 99–100; plectrums, 139; purse frames, 100; puzzle balls, 85–86; puzzles, 165; rings, 109; rulers, 199; screens, 120; seals, 192; snuff bottles, 101; stirrups, 107; sword guards, 184; talismans, 176; utensils, 113, 114, 115, 118; vases, 102; water vessels, 102; writing implements, 188, 189, 190
Chinese Ladder puzzle, 165
Chinese Rods and Cords puzzle, 166
Ching dynasty, 17, 79, 97, 128, 131, 188, 190

chisels, gouger, 201
choppers, 113
chopping blocks, 113
chopsticks, 17, 18, 31, 113; holders for, 18, 113
Chou dynasty, 17, 104, 109, 113, 195
chowrie, 171
Christianity: devotional ivories used in, 21, 77; influences on ivory art, 20, 21; themes of, in ivory art, 22, 90, 98, 166, 167–68, 194
chryselephantine statues, 80
cigar boxes and cigarette cases, 179
cigarette and cigar holders, 17, 46, 179
cigar-tip cutters, 179
clappers. *See* castanets
clock cases, 23, 40, 91–92
clothes brushes, 106
clothes forks and paddles, whalebone, 111
clothespins, 36, 106
Cluny Museum, 115, 171
coasters, 18, 114
cocktail picks, 18, 114
coffers, 18, 98
coins, 197–98; display cases for, 99
cold cream: boxes for, 161; effect on ivory of, 42
collar buttons, 46, 105
collars, artificial ivory, 40
collections, ivory, 27
Colombo Museum, 177
combs, 16, 17, 18, 20, 21, 26, 40, 46, 161–62; liturgical, 172; "nit," 162
compasses, 18; architectural, 177; directional, 178
Congo, 25
containers, 16, 26, 36, 40, 41, 95–103; canes as, 98; eye-paint, 18, 162; for flowers, 99; ointment, 16; salve, 133. *See also under* boxes; cases; coffers; caskets for valuables; inro; vessels
conversion scales, 178
cookie cutters and molds, 114
corkscrews, 195–96
corset stays, baleen, 36
costume jewelry, 30, 39
costume ornaments, 46
cotton spinners, 93

[232] *Index*

couches, 16, 119
counterfeit balls, 87
counterfeit boxes, 22, 86
Crete, 163, 176
cribbage boards, 126
crochet hooks, 93
crosses, 167–68; pectoral, 168
crucifixes, 78, 167–68
cuff links, 106
cups, 114, 177; paintbrush, 190

daggers, 36; handles for, 18, 19, 184–85; sheaths for, 184–85
dance-card booklets, 187–88
darning eggs, 93
de-knotters, 196. *See also* knot unravelers
dentures. *See* artificial teeth
de Pont, Alexander and Sylvius, 40
desk sets, artificial ivory, 40
dice, 16, 18, 20, 36, 38, 46, 126; cups for, 126; poker, 129
Dieppe, 22–23, 87, 95, 125, 158, 200
dies, potters', 18, 192
dippers, 102
diptychs, 90, 168–70, 194
directional compasses, 178
dishes, 114
display cases, coin, 99
doctor's lady, 17, 131
dolls, 176, 181–82
dominoes, 36, 46, 126
door posts, 136
doors, 19, 135; handles and knobs for, 36, 135; panels on, 20
doum-palm nut, 38–39
draughts. *See* checkers
drawer knobs, horn, 36
Dresden, 22
dresser sets, artificial ivory, 40
drinking vessels, 36, 114–15
drums, musical, 18, 137
duplicating machines. *See* reproducing machines
dusters, 202
dye, hair, 163

ear, models of, 22
earrings, 37, 106
ear spoons, 164

East Indies, 37
eating utensils, 115
ecclesiastical staffs, 170–71
Egypt, 16, 19, 77; specific items made in: amulets, 176; arrowheads, 183; bracelets, 103; castanets, 136–37; containers, 96; eating utensils, 115; fan handles, 140; toys, 181; fishhooks, 185; furniture, 119; games and accessories, 124, 126, 128; grooming articles, 161, 163; jewelry, 109; labels, 188–89; spatulas, 200; tools, 92, 93; vases, 101. *See also* Tutankhamen, King, tomb of
electrical insulators, 136
elephant, 23–24; African, 28, 29; Asiatic, 28; game laws concerning, 30; Indian, 28; ivory of, 29–30, 82; teeth of, 30; tusk of, 16, 17, 24, 26, 28–30, 42; tusk of, used for drums, 137; tusk of, used for lampshades, 120
Embriachi family, 21, 108
England, 21, 161; miniature carving in, 23; specific items made in: boxes and cases, 97, 101, 108; dagger handles, 184; horse trappings, 108; oliphants, 198; pipe stoppers, 180; seals, 192–93; scrimshaw, 88; walking sticks, 110
envelope cases, 18, 99
Erbach, 23
erotic art, 79
Eskimo game, 126–27
Eskimos, 26, 32; specific items made by: animal-hide scrapers, 93; boat fittings, 195; brow bands, 104; buckles, 104; buttons, 105; charms, 176; clock cases, 92; cribbage boards, 126; dolls, 176, 182; handles, 197; necklaces, 109; penholders, 192; pigment, 188; pipes, 180; purse fasteners, 100; snow goggles, 110; toggles, 201; tops, 182; tools, 201–2; *ulu* cutter, 113; watchbands, 92; weapons, 183, 184, 185
Essex Institute, 160
Europe, 19–23, 132. *See also* entries for individual countries
eye, models of, 22

Index [233]

eye-paint containers, 18, 162

face-powder boxes, 162
face-powder brushes, 162
Faistenberger, Andreas, 78, 168
fans, 17, 18, 22–23, 140, 157–58; handles for, 18, 140
faucet handles, 136
Faydherbe, Lucas, 114
fencing-foil guards, 18, 184
fertilizer, 196
fetishes, 175–76
fids, 196
Field Museum of Natural History, exhibits in: boxes, 100, 162, 180; combs, 162; dish, 114; fan, 157; gourd with ivory cover, 96; hornbill, 37; massager, 132; narwhal tusk, 32; rulers, 199; toe scratcher, 140; writing implements, 188, 189, 190
figurines, 16, 18, 21, 38, 77, 78
finger rings. See rings
fishhooks, 16, 185
flabella, 171. See also fly whisks
flasks, perfume, 164
flour tryers, 115
flower stands, 102
flutes, 17, 18, 137
fly whisks, 16, 158. See also flabella
fobs, 92
Fogg Museum of Art, 104
food, ivory, 116
footrests, 119
forceps, 160
forks, 115; clothes, 111; handles for, 25; salad-mixing, 117–18
fortune-telling devices, 182
"fossil ivory," 30
fountain pens, artificial ivory, 40
Fowler Museum of Decorative Arts, 101, 176–77, 180, 186
France, 86–87. See also Dieppe
French Ivory, 40; clock cases, 92
furniture, 16, 118–21; ecclesiastical, 21

Gallery of English Costume, 160
gambling tables, 16
game accessories: boards, 16, 127; markers, 16, 128; pieces, 17. See also checkers; chessboards; chess pieces and sets; cribbage boards
gavels, 196
German National Museum, 111
Germany, 22, 23, 79, 87
gingerbread nut. See doum-palm nut
Glenz, Otto, 23
globes, 112–13
glove boxes and cases, 18, 106–7
glove-finger menders, 94
glove-powder boxes, 107
glove stretchers, 107
goblets, 114
goggles, snow, 110
golf clubs, 128
gouger chisels, 201
graters: snuff, 22, 200; tobacco, 180–81
Greece, 19, 20, 139
Guimet Museum, 200
Gyntelberg, Niels, 112

hairbrush backs and handles, 40, 162
hair dye, 163
hairpins, 16, 17, 163; boxes for, 163
hair smoothers, 163
hammers, 201
handbags. See purses
hand-guards, dagger, 184
handkerchief boxes and cases, 18, 99
handles, 16, 18, 26, 34, 36, 197; corkscrew, 195–96; door, 135; fan, 140; faucet, 136; flabella and fly whisk, 158, 171; on grooming articles, 40, 162, 163, 165; on holy-water sprinklers, 171; key-ring, 39; magnifying lens, 91; parasol and umbrella, 23, 159–60; pickwick, 199; toothbrush, 40, 133–34; on utensils, 25, 115, 116, 117; walking stick, 110; water-dipper, 102; on weapons, 18, 19, 31, 184–85; whiskbroom, 202
hand rests, 188
Han dynasty, 17, 192, 195
harpoon heads, 185
harpoon rests, 185
harps, 137
Harrick, Christof, 79
hatchets, 201
hatpins, 107

hats, 107; brushes for, 107; perimeter frames for, 36; stands for, 107
Hawkins, John, 46
headrests, 119–20
head-scratch pins, 140
Herzog Anton Ulrich Museum, 174
high-relief carving, 78
hilts, dagger, 31
hippopotamus: ivory of, 33–34, 82, 130, 178; teeth (tusks) of, 17, 125, 135
History of Medicine Museum, 112
Holtzapffel, Charles, 23
holy-water sprinklers, 171
holy-water vessels, 21, 176–77
hookah stands, 179
hoopskirt hem-frames, baleen, 36
horn, 36–37
hornbill, 37, 48; specific items of: abstinence plaques, 167; snuff bottles, 101
Horniman Museum, 137
horns: drinking, 198; hunting, 19, 21, 25, 185–86; musical, 137–38; powder, 18, 19, 39, 186. See also oliphants
horse trappings, 107–8
houses: of ivory, 135; miniature, 17
howdahs, 18, 197
human figurines, 17
hunting horns. See under horns,
hunting implements, 26, 183–87
Hyatt, John, 39–40
hydrochloric acid. See muriatic acid

Iceland, 31
ice picks, 201
incense boxes and cases, 17, 99
incense burners, 99
incense powder, 172
India, 16, 17, 18–19, 41, 52, 77, 79; specific items made in: alphabet sets, 111; bangles, 103–4; baskets, 103; beads, 109; book covers, 90; calling-card cases, 97; games and accessories, 124, 125, 126, 129; grooming articles, 161, 162, 164; hookah stands, 179; howdah, 197; medicine, 132; palanquins, 198; talismans, 176; vases, 102; watchstands, 92
india ink, 188

Indians, North American, 93
ink-blotter holders, 188
ink-pens, architect's, 187
ink scrapers, 188
inkstands, 18
inkwells, 16, 19, 188
inlay: horn, 36; ivory, 15, 16, 17, 19, 20, 21, 119
inro, 37, 133, 193
insulators, electrical, 136
interior carvings, 86, 87
intra-uterine devices, 131
Islam, 19, 98, 108, 188
Italy, 23, 105
I.U.D. See intra-uterine devices
Ivorine, 40, 81, 92
Ivoroid, 40, 92
ivory: aging of, fake, 50; aging of, natural, 49, 50, 51; care of, 46, 53; carving of, 18, 42–43, 47–48; cleaning of, 53–54, 55; coloration of, 49, 50; coloration of, artificial, 83; coloration of, artistic, 51–52; coloration of, natural, 49; cracked, 53–55; cutting of, 41–42, 45–46; dating of, 50–51; definition of, 28; display of, 53; figures drawn on, 77; as food, 116; hardness of, 48, 59; heating of, 49–50; mechanical cutting of, 45–47; as medicine, 19, 33, 131–32; paintings on, 43; physical properties of, 29–30; preparation of, 41–42; reflective properties of, 58–59; restoration of old, 53, 55; sheet, 43–45, 46; softening of, 43–45; sources of, 28–34; specific gravity of, 29, 56–57; staining of, 52; substitutes for, 35–40; superstitions regarding, 31, 33, 109, 177, 184; tests for, 56–60; trade in, 22, 23–25; use of other materials with, 80–81; visual properties of, 59–60. See also "fossil ivory"
ivory black, 163, 188
Ivory Coast, 23, 25
ivory dust, 15, 18, 33, 132, 161, 163, 172, 188, 196, 197, 200
Ivory House (Zanzibar), 24
ivory jelly, 116
ivory-palm nut, 38

Index [235]

ivory powder. *See* ivory dust

jackstraws, 128
jagging wheels, 117
Janus-head, 25, 79
Japan, 18, 37, 52; specific items made in: buttons, 105; candle holders, 98; daggers, 184-85; ear spoons, 164; fans, 140, 157; latch pegs, 197; medicine, 132; pipe cases, 180; plectrums, 139; seals, 192; shrines, 175; snuff bottles, 101; utensils, 113, 117, 118. *See also* netsuke
jars: lids for, 16, 116; opium, 180; tobacco, 180
Jenkins, Frank L., 23
jeweler's polish, 197
jewelry, 22, 23, 108-9. *See also* costume jewelry
jewelry boxes and cases, 17, 18, 19, 21, 108
jewelry stands and trees, 108
Jewish Museum, 91
Johnson Whaling Collection, 97
Joseph Mayer Museum, 173, 175, 195
"Judgment of Solomon" (statue), 78

Kashmir, 164
Kestner Museum, 112
key-ring handles, betel nut, 39
keys, piano and organ, 26, 46, 138-39
knitting needles, 93
knives, 16, 115; handles for, 17, 21, 26, 115, 199; rests for, 116; snow, 193; snuff, 200. *See also* pocketknife handles
knobs, 16, 39; cabinet, 119; door, 36, 135; drawer, 36; organ, 138; scroll-roller, 91; umbrella-handle, 38; walking-stick handle, 38
knot unravelers, 17. *See also* de-knotters
Kuan Yin, statue of, 79

labels, 16, 188-89
labrets, 108
ladles, 116
lampshades, 41, 120
lanterns, 98
latch pegs, 197

Le Marchand, David, 78
letter openers, 18, 36, 40, 46, 189
Li, Hsao-yu, 86
liturgical combs, 172
lockets, 108, 172
London Museum, 160
Los Angeles County Museum of Natural History, 88, 131, 138
Louvre Museum, 108, 136, 137, 161, 164, 169, 172, 173
Lucas, Richard, 112
lures, fishing, 185
lutes, 138
lyres, 16

machines used in ivory craft. *See* mechanical craftsmanship
Madonna, 22
Madras, 18
magnifying lens handles and rings, 91
mah jong pieces, 17, 128
Malaya, 37, 132
mallets, 202
mammoth: ivory of, 30, 82, 188; tusk of, 21, 26
manicure intruments, 163
maps, 112-13
marbles, 16, 128
Margret, the priest's wife, 31
marrow-spoon handles, 116
massagers, 132
mastodon ivory, 30
match safes, 179-80
measuring sticks, 94. *See also* rulers
mechanical craftsmanship, 22, 23, 42-43, 86
medallions, 16
medals, 197
medical instrument cases, 132-33
medicine, 19, 33, 131-32; cases for, 133
Megiddo excavations, 16, 124, 137, 161, 192
memento mori, 79
memorandum pads and tablets, 189
memorandum slips, 97
menu cards, 116
metronomes, 92
Metropolitan Museum of Art, 17, 180, 190, 196

Middle East, 78
Miller, Leonhart, 91
Milwaukee Public Museum, 80
Ming dynasty, 17, 189
miniatures, 23, 79, 80, 197
mirror parts and accessories: cases, 21, 22, 164; frames, 23, 40; handles, 16, 163–64
models: anatomical, 111–12; architectural, 112
molds, cookie, 114
money, 197–98; cases for, 18, 99
mortar and pestle. See pestle-and-mortar sets
Moscow, 21
Moslem beliefs regarding ivory, 31, 132. See also Islam
mouthpieces, musical instrument, 137
muriatic acid, effect on ivory of, 43–44
Museum of Decorative Arts, 79, 112
Museum of Fine Arts, 85, 137
Museum of the City of New York, 160
Museum of the History of Science, 177
musical instruments, 20, 21, 136–39. See also entries for individual musical instruments
Myrick, Frederick, 88
Mystic Seaport Museum, exhibits in: birdcage, 97; cane, 131; clock case, 92; clothes paddle, 111; clothespins, 106; duster handle, 202; fids, 196; mallet, 202; pickwicks, 199; rattle and whistle, 182; rope-bedspring tighteners, 120; seam rubbers, 94; shawl clips, 109; swifts, 95; truncheons, 187; utensils, 116, 117

nameplates, 46, 198–99
Nantucket Historical Association, 88, 89, 105
napkin rings, 32, 116
narwhal: ivory of, 32–33, 82, 132; tusk of, 32–33, 121, 131, 132, 176, 184, 200
National Maritime Museum, 178
National Museum (Copenhagen), 112
National Museum (Florence), 86, 164, 171
National Museum of the Society of Antiquaries, 132

navettes, 22, 95
navigational instruments, 178
neck dusters, 160
neckerchief slides, 109
necklaces, 16, 109
needle cases, 18, 38, 94
needles: sewing, 94, 95; threading, 95; yarn, 95
net shuttles, 94
netsuke, 18, 32, 36, 79, 80, 81–85
Newark Museum, 85
New Bedford Whaling Museum, 89
Nimrud excavations. See Assyria
"nit combs," 162
nude figurines, 16. See also doctor's lady

ointment containers, 16
ojime, 37, 82
okimono, 82, 84
Old Chateau Museum, 91
Old Dartmouth Historical Society, 117
Old Town Museum, 91
oliphants, 174, 198
opium jars and pipes, 180
orange peelers, 18, 116–17
organ-key veneers, 46, 138–39
organ knobs and stops, 138
ornaments, 17, 26, 37, 38

pagodas, miniature, 17, 79
pails, 17, 102
paintbrush parts and accessories: cases, 17, 190; cups, 190; handles, 17
painting brushes, 190
paintings on ivory, 43
Palace Art, 102
palanquins, 198
palettes, 190
panels, 16, 23, 77, 120
Pape, H., 45
paper creasers, 36, 190–91
paper rougheners, 191
paperweights, 18, 191
parasol handles, 23, 159–60
Paris, 21
Parthenon, models of, 112
pastry cutters, 117
pastry-decorator syringes, 117
paxes, 172–73

pencils, 97, 191; caps for, 97; trays for, 191
pendants. *See* lockets
penholders, 18, 36, 192
pens, 191–92; architect's ink-, 187; cases for, 16, 191–92; fountain, 40; racks for, 18; trays for, 191
perfume flasks and perfumers, 22, 164
Persia, 31, 129
personal groomer sets, 164
pestle-and-mortar sets, 117
Phidias, 19, 43, 45, 77
Phoenicia, 16, 77, 103, 161
phosphoric acid, effects on ivory of, 44
piano-key veneers, 26, 46, 138–39
pianos, 45
piccolos, 137
picks. *See* plectrums
"pick-up-sticks." *See* jackstraws
pickwicks, 199
picture frames, 18, 40, 99–100
pie crimpers, 117
pill boxes, 18, 133
pins, 16, 17, 199; basting-thread removal, 93; boxes for, 100. *See also* hatpins
pipes, 180; cases for, 18, 180; stoppers for, 180
pistols, 186
pitchers, 102
Pitt Rivers Museum, 200
planetariums, 178
Planzone, Filippo, 86
plaques, 16, 79, 90; abstinence, 31, 167; religious, 170; wall, 121
playing cards, 129
playing pieces, 16
plectrums, 17, 139
pocketknife handles, 199
pointers, 113; reading, 91
poker chips, 38, 129
poker dice, 129
polish, 197
polyptychs, 169
porpoise, jawbone of, 35
portable kits, ladies', 197
portrait medallions, 78
portraits painted on ivory, 43, 78
Portuguese trade in ivory and slaves, 22, 24

potters' dies, 18, 192
pounce pots, 192
powder horns. *See under* horns
Pratt(-Read) Company, 45, 46
prayer beads, 174–75
prayer wheels, 173
purses, 100; fasteners for, 100; frames for, 100
puzzle balls, 17, 85–87, 125
puzzle boxes, 165
puzzles, 165–66
Pyralin, 40
pyxes, 20, 21, 173

rankets, 139
rattles, 182. *See also* shaker instruments
razors, 165
reading pointers, 91
Regensburg, 22
religious themes in and uses of ivory, 21, 77, 166–67, 198. *See also* Buddhism; Christianity; Islam
reliquaries, 21, 173–74
Reliquary Museum, 174
Rennaissance, 24
replicas, 197
reproducing machines, 23, 26, 46–47
retables, 21, 174
Revere, Paul, 187
rhinoceros, horn of, 36
riding whips, 199
rifles, 186
rings, 16, 17, 21, 31, 37, 108, 109, 132, 175
rollers, scroll, 91
rolling pins, 117
Rome, 20–21, 96, 108, 161, 194, 195; Eastern Empire, 21
room screens, 120
rope-bedspring tighteners, 120
rope twisters, 199
rosaries, 174–75
roulette-wheel balls, 129
Royal Academy, 47
Royal Society of Miniature Artists, 23, 47
rulers, 17, 18, 199. *See also* measuring sticks
Russia, 21, 102

saddles, 20, 108
St. Petersburg, 21
salad utensils, 117–18
salt-and-pepper shaker sets, 118
salt cellars, 25, 118
salve containers, 133
sandals, 109
Santa Barbara Historical Society Museum, 182, 202
scales, conversion, 178
Scandinavia, 31
scent bottles, 18
scent boxes, 17, 100
scepters, 20, 199–200
Schatzkammer Museum, 185–86
Schnütgen Museum, 161
Science Museum (London), 47
scientific instruments, 177–78
scimitar handles, 19
scissors, 202
scoops, 116; tea, 118
scoring devices, 129
screens: folding, 17; room, 120
screws, 203
scrimshaw, 27, 34, 48, 87–89
scrivelloes, 29
scroll rollers, 91
seal and seal-pigment boxes, 17, 193
seal attracters, 186
seals, 17, 192–93
seam rubbers, 94
sea shells, 87
Seattle Art Museum, 37
secret compartments in ivory objects, 96, 160
servant's call signal, 136
sewing awls. See bodkins
sewing birds, 94
sewing needles, 92, 94, 95
shaker instruments, 139
Shang dynasty, 17, 102, 115, 163
shaving-brush handles, 40, 165
shawl clips, 109
sheet ivory. See under ivory
shields, 29, 186
ships: adorned with ivory, 77; models of, 26
Shiva, statues of, 79
shoehorns, 18, 36, 109–10

shrines, 175; replicas of, 18
Siam, 132, 176
Siberia, 30
Singapore ball, 79, 87
size, 200
skeletons carved in ivory, 79, 112
skulls, 82. See also memento mori
slave trade, 22, 24, 25
sled runners, 26, 200
slide rules, 177–78
slipper clips, 110
Smilis, 19
snow goggles, 110
snow knives, 193
snow-shovel edges, 202
snuff accessories: bottles, 17, 37, 101; graters, 22, 200; knives, 200; saucers, 101; snuffboxes, 40, 100–101
soap dishes, artificial ivory, 40
Soma, Sen-rei, 18
Southeast Asia, 17
Spain, 98, 140, 157, 161, 201
spatulas, 16, 117, 200–201
spear points, 183
spear rests, 185
sperm-whale teeth, 34, 82, 87, 181, 188. See also whale-tooth ivory
spike balls, 87
spill cases, 101
spindles, 95
spinning wheels, 93
Spitzer Collection, 27
spools, thread, 93, 95
spoons, 16, 36, 115; handles for, 25, 116; salad-mixing, 117–18; with snuff bottles, 101; strainer, 118; water, 193
Sri Lanka, 28. See also Ceylon
staffs, 21, 170–71
stamp boxes and cases, 18, 101
stands: hat, 107; hatpin, 107; hookah, 179; jewelry, 108
statues, 19, 43, 45, 77, 78, 80
statuettes, 16, 77, 78
Stenbock, Gustav Magnus, 86
Stephany and Dresch, 23
stirrups, 107
stoups, 177
strainer spoons, 118
Sudan, 25

sundials, 91
Sung dynasty, 91, 99, 193
superstitions. *See under* ivory
surgical instruments, 133
Su-shen tribesmen, 17, 183
swifts, 94–95
switches, electrical, 136
sword parts and accessories: handles, 184; hilts, 17, 36; sheaths, 18
Syria, 16, 119
syringes, pastry-decorator, 117

tables, 16, 19, 120; gambling, 16
tablets: memorandum, 189; writing, 17, 21, 189, 193–94
talismans, 18, 175–76
Tang dynasty, 17, 108, 137, 139, 189, 199
tangrams, 90, 166
tankards, 114
tatting shuttles, 95
tea accessories: caddies, 118; scoops, 118; teapots, 118
tea houses, miniature, 79
teapoys, 18, 120
teeth. *See under* elephant; hippopotamus; sperm-whale teeth; whale-tooth ivory
teethers and teething rings, 131
telescopes, 178
temples, miniature, 17, 79
Teuber family, 22
Thailand. *See* Siam
thimbles, 38, 93, 95; cases for, 95
thread barrels, 95
threading needles, 95
thread spools, 93, 95
thrones, 18, 20, 121
thumb-guards, archers', 18, 183
thumb-rings, archers', 37
tickets, 195
Tippo Tib, 25
tobacco accessories: boxes, 180; graters, 180–81; jars, 180
toe scratchers, 140
toggles, 201
toiletry articles, 19
Tokyo National Museum, 85
tongs, salad, 118

tongue depressors, 133
tongue scrapers, 133
tools: for carving, 18, 42, 47, 89; of ivory, 26, 32, 201–2
toothbrush handles, 40, 133–34
tooth-cleaning powder boxes, 134
toothpicks, 46, 134; cases and holders for, 134
tops, 182; gambling, 129–30; game, 130
toys, 26, 181–82
trade. *See under* ivory; slave trade
trays, 118; pen-and-pencil, 191
tremolos, organ, 138
trinity rings, 22, 109
triptychs, 169
Troger, Simon, 78
trumpets, 137–38
truncheons, 186–87
trunk framework, baleen, 36
tusk. *See under* elephant; hippopotamus; mammoth; narwhal; walrus
tusks, carved, 25
Tutankhamen, King, tomb of, 103, 108, 110, 119, 127, 187, 190
tweezers, 160

Ulrich Museum, 185
ulu cutter, 113
umbrellas, 159–60
United States, 26
United States National Museum, 182, 185
University of Alaska Museum, 185
University of Pennsylvania Museum, 37, 180
utensils, 26, 115

vases, 16, 22, 101–2
Vatican Museum, 90, 167, 172
vegetable ivory, 38–39, 48, 59
veneer, ivory as, 45, 46, 138–39, 186
Vereschagin, N.S., 102
vessels: drinking, 102, 114–15; religious, 176–77
Victoria and Albert Museum, 23, 78; exhibits in: books, 90; book cover, 90, 167; caskets, 99; comb, 161; dagger handle, 184; fan, 157; furniture, 119; hunting horn, 186; mirror

cases, 164; musical instruments, 136, 137; netsuke, 85; oliphants, 198; religious articles, 167, 168, 169, 171, 173, 177; snuff graters, 200; table, 120; umbrella handles, 160; utensils, 114, 115; water-dipper handle, 102
Vikings, 32–33, 131, 176, 184, 200
Villerme, Joseph, 168
vinaigrettes, 134

Wales, 116
walking sticks, 18, 110. *See also* canes
Wallace Collection, 183, 186
wall plaques, 121
walrus, tusk of, 16–17, 26, 31, 87, 88
walrus ivory, 17, 31–32, 82, 132; specific items of: armor, 184; artificial teeth, 130; back scratchers, 140; boat fittings, 195; candelabra, 98; cribbage boards, 126; fishhooks, 185; handbag fasteners, 100; handles, 197; pipes, 180; religious articles, 167, 168; rolling-pin handles, 117; snow goggles, 110; toggles, 201; tools, 201–2; toothbrushes, 134; toys, 182
warthog, tusk of, 34
watch accessories: stands, 18, 92; watchbands, 92; winding keys, 92
water, effect on ivory of, 42
water dippers, 18
water spoons, 193
water vessels, 102. *See also* vessels
Watt, James, 46
weapons, 183–87
wedges, 202
whale. *See* baleen whale; whalebone; whale-tooth ivory

whalebone, 34, 35, 36, 47–48; specific items of: alphabet sets, 111; canes, 131; clothing accessories, 106, 110, 111; doorknobs, 135; fids, 196; lanterns, 98; mallets, 202; measuring sticks, 94; pipe stoppers, 180; pointers, 113; rope-bedspring tighteners, 120; rope twisters, 199; scrimshaw, 87; seam rubbers, 94; swifts, 95; toys, 181; umbrella ribbing, 159; utensils, 113, 115, 116, 117; weapons, 184, 187; whisk brooms, 202
whalemen. *See* American whaler seamen
whale-tooth ivory, specific items of: jewelry trees, 108; pickwicks, 199; pounce pots, 192; seals, 193; utensils, 114, 115, 117, 118. *See also* sperm-whale teeth
Whaling Museum, 98, 105
whips, 187; riding, 199
whisk brooms, 202
whistles, 182, 202
window-shade pulls, 121
wrist guards, archers', 183
writing, microscopic, on ivory, 79
writing accessories, 17; brushes, 190; styli, 193; tablets, 17, 21, 189, 193–94

Xylonite, 40

yardsticks, 94
yarn needles, 95

Zanzibar Island, 23, 24
Zeus, statue of, 19, 43, 45
Zick family, 22; Lorenz, 22, 86; Stephan, 22, 86, 109, 111, 112